T5-BBB-466

INTEGRATIVE FACILITIES MANAGEMENT

THE IRWIN/APICS LIBRARY OF INTEGRATED RESOURCE MANAGEMENT

Integration Functions

Managing Information: How Information Systems Impact Organizational Strategy
Gordon B. Davis and Thomas R. Hoffman

Managing Human Resources: Integrating People and Business Strategy *Lloyd Baird*

Managing for Quality: Integrating Quality and Business Strategy *V. Daniel Hunt*

World-Class Accounting and Finance *Carol J. McNair*

Customers and Products

Marketing for the Manufacturer *J. Paul Peter*

Field Service Management: An Integrated Approach to Increasing Customer Satisfaction
Arthur V. Hill

Effective Product Design and Development: How to Cut Lead Time and Increase Customer Satisfaction *Stephen R. Rosenthal*

Logistics

Integrated Production and Inventory Management: Revitalizing the Manufacturing Enterprise *Thomas E. Vollmann, William L. Berry, and D. Clay Whybark*

Purchasing: Continued Improvement through Integration *Joseph Carter*

Integrated Distribution Management: Competing on Customer Service, Time, and Cost *Christopher Gopal and Harold Cypress*

Manufacturing Process

Integrative Facilities Management *John M. Burnham*

Integrated Process Design and Development *Dan L. Shunk*

Integrative Manufacturing: Transforming the Organization through People, Process, and Technology *Scott Flaig*

INTEGRATIVE FACILITIES MANAGEMENT

John M. Burnham

IRWIN
Professional Publishing

Burr Ridge, Illinois
New York, New York

This publication is designed to provide accurate and authoritative information in regard to the subject matter covered. It is sold with the understanding that neither the author nor the publisher is engaged in rendering legal, accounting, or other professional service. If legal advice or other expert assistance is required, the services of a competent professional person should be sought.

From a Declaration of Principles jointly adopted by a Committee of the American Bar Association and a Committee of Publishers.

Editor-in-chief: Jeffrey A. Krames
Project editor: Jean Lou Hess
Production manager: Ann Cassady
Designer: Larry J. Cope
Compositor: Carlisle Communications, Ltd.
Typeface: 11/13 Times Roman
Printer: Arcata Graphics/Kingsport

Library of Congress Cataloging-in-Publication Data

Burnham, John M.
Integrative facilities management / John Burnham.
p. cm. — (The Business One Irwin/APICS library of integrated resource management)
Includes bibliographical references and index.
ISBN 1-55623-679-4
1. Facility management. 2. Factory management. I. Title.
II. Series.
TS177.B87 1994
658.2—dc20 93-14481

Printed in the United States of America

1 2 3 4 5 6 7 8 9 0 AGK 0 9 8 7 6 5 4 3

FOREWORD

Integrative Facilities Management is one book in a series that addresses the most critical issue facing manufacturing companies today: integration—the identification and solution of problems that cross organizational and company boundaries—and, perhaps more importantly, the continuous search for ways to solve these problems faster and more effectively. The genesis for the series is the commitment to integration made by the American Production and Inventory Control Society (APICS). I attended several brainstorming sessions a few years ago in which the primary topic of discussion was, "What jobs will exist in manufacturing companies in the future—not at the very top of the enterprise and not at the bottom, but in between?" The prognostications included:

- The absolute number of jobs will decrease, as will the layers of management. Manufacturing organizations will adopt flatter organizational forms with less emphasis on hierarchy and less distinction between white collars and blue collars.
- Functional "silos" will become obsolete. The classical functions of marketing, manufacturing, engineering, finance, and personnel will be less important in defining work. More people will take on "project" work focused on continuous improvement of one kind or another.
- Fundamental restructuring, meaning much more than financial restructuring, will become a way of life in manufacturing enterprises. The primary focal points will be a new market-driven emphasis on creating value with customers, as well as greatly increased flexibility, a new business-driven attack on global mar-

kets which includes new deployment of information technology, and fundamentally new jobs.

- Work will become much more integrated in its orientation. The payoffs will increasingly be made through connections across organizational and company boundaries. Included are customer and vendor partnerships, with an overall focus on improving the value-added chain.
- New measurements that focus on the new strategic directions will be required. Metrics will be developed, similar to the cost of quality metric, that incorporate the most important dimensions of the environment. Similar metrics and semantics will be developed to support the new uses of information technology.
- New "people management" approaches will be developed. Teamwork will be critical to organizational success. Human resource management will become less of a "staff" function and more closely integrated with the basic work.

Many of these prognostications are already a reality. APICS has made the commitment to *leading* the way in all of these change areas. The decision was both courageous and intelligent. There is no future for a professional society not committed to leading-edge education for its members. Based on the Society's past experience with the Certification in Production and Inventory Management (CPIM) program, the natural thrust of APICS was to develop a new certification program focusing on integration. The result, Certification in Integrated Resource Management (CIRM) is a program composed of 13 building-block areas that have been combined into four examination modules, as follows:

Customers and products
  Marketing and sales
  Field service
  Product design and development

Manufacturing processes
  Industrial facilities management
  Process design and development
  Manufacturing (production)

Logistics
  Production and inventory control
  Procurement
  Distribution

Support functions
- Total quality management
- Human resources
- Finance and accounting
- Information systems

As can be seen from this topical list, one objective in the CIRM program is to develop educational breadth. Managers increasingly *must* know the underlying basics in each area of the business: who are the people who work there, what are day-to-day *and* strategic problems, what is state-of-the-art practice, what are the expected improvement areas, and what is happening with technology? This basic breadth of knowledge is an absolute prerequisite to understanding the potential linkages and joint improvements.

But it is the linkages, relationships, and integration that are even more important. Each examination devotes approximately 40 percent of the questions to the connections *among* the 13 building-block areas. In fact, after a candidate has successfully completed the four examination modules, he or she must take a fifth examination (Integrated Enterprise Management), which focuses solely on the interrelationships among all functional areas of an enterprise.

The CIRM program has been the most exciting activity on which I have worked in a professional organization. Increasingly, manufacturing companies face the alternative of either proactive restructuring to deal with today's competitive realities or just sliding away—giving up market share and industry leadership. Education must play a key role in making the necessary changes. People working in manufacturing companies need to learn many new things and "unlearn" many old ones.

There were very limited educational materials available to support CIRM. There were textbooks in which basic concepts were covered and bits and pieces that dealt with integration, but there simply was no coordinated set of materials available for this program. That has been the job of the CIRM series authors, and it has been my distinct pleasure as series editor to help develop the ideas and facilitate our joint learning. All of us have learned a great deal, and I am delighted with every book in the series. But the spirit of continuous improvement is built into the CIRM program and into the book series.

Thomas E. Vollmann
Series Editor

PREFACE

The operational manufacturing or distribution facility, corporate headquarters or sales office, and all associated real estate and other improvement represent a triumph of discipline over all the disruptions that beset most projects on their way to realization. Unfortunately, this triumph often seems to go unheralded in the rush to "get on with it."

Besides being a marvel by virtue of working at all, the modern facility is a very visible sign of a company's corporate strategy, and worthy of considerable study. In fact, astute competitors do just that since the facility size, construction timing, and product assignments describe how the company proposes to position itself in the marketplace.

However, there is not much hard evidence of the kind of careful, thoughtful planning in the development of industrial facilities that routinely accompany new products, new equipment, or even new offices. Many sites "grow like Topsy," seeming to *occur* rather than being planned at all. Attention should be called to the very great benefits that can be passed over if a plant is just "thrown together."

The typical industrial facility is created like the traditional product design, in the development ivory tower, and then tossed over the wall for execution by manufacturing—somehow. The length of the development cycle—time to market—is made even longer by the many phases of rework, engineering changes, and so forth. The sordid history of product development is often surpassed in ineptness by the frequently haphazard approach to the industrial facility. Moreover, as contrasted with the products it manufactures, which may be short-lived, the facility itself can be expected to be active for 20 to 30 years and deserves, therefore, several times the effort accorded to any product. Unfortunately, this often doesn't happen.

The industrial facility, like manufacturing itself, has for decades been the unequal workhorse on the corporate team. Yet high-quality, competitively priced products generate the profits that keep the "engine of industry" running. As the text will show, the facility and its working "contents" can easily make the difference between profit and loss in areas like quality (equipment capability), materials handling and layout (flexibility, short throughput times), ease of modification, and the management of by-products and regulatory mandates. What's needed is IFM—integrative facilities management.

Management teams, simultaneous-engineering teams, problem-solving teams, small-group improvement teams, and self-managed work teams have become part of the working vocabulary of those progressive company managers who seek continuous improvement and low-waste production and distribution. This text proposes another team: the site development team—to follow the "facility product" from its birth as an idea into commissioning and then into ongoing operation. This team, like the product-process development and other management teams, will be made up of multifunctional and multilevel people focused on delivering the right facility at the right time with the right capacity, equipment, and cost.

The rationale for doing this is simple: we can't afford *not* to do it! The strong showing that Japanese products have made around the globe emphasizes the benefits of teamwork and concentration on flawless execution. Many Japanese plants are spartan, but *they work*. A recent survey noted that the United States spends more than a trillion dollars a year on manufacturing facilities and multifamily construction. With a 12-year oversupply of commercial office space, it is anticipated that most of these funds go into industrial facilities. And if new industrial facilities can include many of the ideas set forth in this text, the total cost of products will be minimized over the useful decades of a facility's life—surely a worthy objective.

A Background of Learning

The writing of *Integrative Facilities Management* was based in large part on years of experience as an operations manager, engineer and designer, and contracts administrator with Texaco's Marine Department; the U.S. Department of Commerce, Maritime Administration; and Pan American World Airways, Aerospace Services Division.

There were countless opportunities to learn from mistakes, in becoming competent in ship operations and working with the facility (engineering plant, deck layout, cargo gear, and so forth). The same was true when I was later part of a facilities design team with the Maritime Administration's Office of Ship Construction (Washington, DC), where we examined the economics of meeting the needs of ship owners and operators, as well as those of the customers for the services the ships were to provide. As a systems engineer for Pan American at Cape Kennedy, it became necessary to view ships in a rather different way—as operating platforms for instrumentation in support of the space programs Mercury, Gemini, and Apollo.

Since 1970 my operations experience has been gained in more traditional settings: airlines, at the University of Miami, Florida; petrochemicals and other petroleum-related activities, at the University of Texas of the Permian Basin (Odessa); a wide range of manufacturing in small-scale production facilities in Valencia, Venezuela; and in many regional manufacturing plants in the Southeast, while based at Tennessee Technological University (Cookeville). Each has heightened my appreciation of the interactions and conflicts between various functional experts who, together, accomplish the entirety of the manufacturing task.

THE PLAN OF THE BOOK

Text development will follow a three-horizon approach, outlined in this simplified table showing integrative facilities management hierarchies and roles.

Role	*External*	*Internal*	*Manufacturing*
Strategic "Representing"	Customers Competition Environment	Capacity Location Resources	Key Tasks Competences Mission
Tactical "Enabling"	Markets Site development Flexibilty	Contingencies Modularity Phasing	Scope Products Responsiveness Manufacturing concept
Operational "Supporting"	Dynamics Environment Constituencies	Adaptive Safety, By-products LLCC	Continuous improvement Capability Site support

Section I: Strategy and Representation

Chapter 1 provides an introduction to IFM representation and strategy, enabling tactics, and supporting operations, placed in a multihorizon, multifunctional context.

Chapter 2 presents conflict situations inherent in multifunctional relations. Jargon, differing performance criteria, and vastly different points of view present potential barriers to achieving consensus. In order for IFM and the site team to be effective, these issues must be dealt with during the various levels of planning and operations support. Readers may use this chapter as a checklist for examining potential challenges to achieving smooth support and planning.

Chapter 3 relates IFM to the role of the integrated resources manager. This entails the detailing of strategic IFM concerns and those of various other corporate and mid-management functional specialists. Capacity and facility strategy details are explained and individual facility strategy elements are developed, leading to the start-up (greenfield) plant mission and charter. Examples will include companies with very high value products, like Hewlett-Packard, and those with relatively low value products, like Nucor and ALCOA.

Section II: Enabling

Chapter 4 addresses the overall activities of IFM in facilities planning, especially in trade-offs assessments and contingency planning. This planning converts the site's mission and greenfield charter into a framework for later detailing as engineering plans and specifications. Many interfaces with other functional areas—especially logistics, engineering, and finance—make this phase of IFM critical to overall plant success in operation.

Chapter 5 makes extensive use of manufacturing system analysis, tying together capacity and flow rates to study salable outputs, required resources inputs, and wastes or by-products that must be dealt with as a portion of the IFM activity. This is especially critical in the proactive IFM roles for both the internal manufacturing customer and the external community and regulatory constituencies.

Chapter 6 examines the nitty-gritty of facilities planning: design, structure, layout, materials flow and handling, and the detailed planning for various support functions assigned to the site development group.

Chapter 7 moves toward implementation of the facilities plan, through project planning and management of the construction program that will convert the industrial facility from plan into reality. The turnover from project team to manufacturing through the commissioning and startup activities brings the technical phase to a pause, while operations begin.

Section III: Operations and Support

Chapter 8 addresses maintenance in all three modes: strategy, enabling tactics, and support. Proactive maintenance is presented as the means through which the plant's process capability can be established and supported throughout production.

Chapter 9 shows another aspect of IFM activities—those which support manufacturing operations. The focus of the site development team is carrying out the plans developed earlier, modified to suit actual experiences during production. The many internal and external factors considered during planning must be effectively addressed, the manufacturing activities supported, and the plant kept secure from liability, regulatory, and health complaints. A large number of details must be kept in mind while addressing the ever changing details of the operating plant.

Chapter 10 addresses the needs of internal and external customers. The issues of unmet expectations (what was demanded but not provided) and ways to plan for satisfying these customers are emphasized.

GOALS

The book carries the reader through the entire process of industrial facility design, execution, and maintenance. Topics move from corporate strategy through to the individual site plan, and from aggregates to very detailed guidance for the integrated resources manager. The directive treatment of strategy as an integral part of the facility planning process and the inclusion of systems analysis, contingency planning, and project management as parts of the necessary tool kit are statements of what truly *must* occur for good results. Because even deathless prose is hard to absorb in large doses, we provide many brief examples and mini-cases throughout the text, as a kind of leaven in the loaf.

Like all the other books in the APICS/CIRM series, the subjects of conflicts, integrative needs, and the nature of resolution processes are given considerable emphasis, especially in Chapter 2. Some detailed case studies are included in chapter appendixes.

ACKNOWLEDGMENTS

The development of an integrative text on facilities management involves the literature of many functional specialties: corporate strategy, manufacturing policy, facilities planning, systems analysis, project management and PERT/CPM, product and process design and manufacturing engineering, maintenance engineering, team building, human resources management, quality, productivity, and, of course, the large body of materials on production planning and control. Many of the sources are noted in the Reference lists that follow most chapters.

Some of the contributions to the writing of this book were so significant that I must single them out with sincere thanks. Professor Wickham Skinner started many academics and practitioners on the upward path with "Manufacturing—Missing Link in Corporate Strategy" (*Harvard Business Review,* 1969) and *Manufacturing in the Corporate Strategy* (New York: John Wiley & Sons, 1978). James A. Tompkins and William A. White—both outstanding industrial engineers, researchers, and consultants—helped in many ways through their classic *Facilities Planning* (New York: John Wiley & Sons, 1984), and they have granted permission to use a number of their figures and tables. Professors William H. Hayes and Steven C. Wheelwright, authors of many articles and texts, the latest being *Regaining Industrial Competitiveness* (New York: John Wiley & Sons, 1988), helped with their ideas of capacity and facilities strategies. Professor Joseph J. Moder, a friend since my University of Miami days, quite literally wrote the book on network methods for project management. Colonel Lane C. Kendall (USMC Ret.), head of the Department of Ship Management, United States Merchant Marine Academy, many years ago when I was a cadet-midshipman there, has been a mentor, challenger, and friend for over 40 years. The sixth edition of his *The Business of Shipping* offered another integrative framework for this text.

Dr. Robert R. Bell and I co-authored a text, *Managing Productivity and Change* (Cincinnati: South-Western Publishing Co, 1991), that was in its own way an integrative work. A number of the figures in *Integrative Facilities Management* are drawn from our book, with Bob's and the publisher's kind permission.

Another learning experience recently helped to crystallize my feelings about the strategic nature of the facilities management challenge. This occurred when Dr. Ramachandran "Nat" Natarajan, who is on the

Tennessee Technological University faculty, was developing a study course for manufacturing processes for the APICS/CIRM program. The module included facilities management, manufacturing processes design, and manufacturing operations. Nat asked me to take on the course overview, facilities strategy and planning, facilities management and forward planning, internal relationships, and external relationships, while he handled manufacturing processes design strategies, manufacturing process design tactics, manufacturing concepts, and manufacturing operations. This text uses a good deal of the background material from my session drafts and some of Nat's ideas about facilities. His criticisms, and the improvements resulting from them, are very much appreciated.

APICS supported us in the development of the manufacturing processes study course and, by additional grants, in my creation of the IFM text. APICS has given permission to use some of the figures and other draft materials from that course, expanded in scope and containing detail to meet the needs of professionals in the facilities management field and those studying for CIRM.

Tennessee Technological University (TTU) contributed my office space and the use of a personal computer to do my part of the study course. Nat and I had the help of Ms. Jody Norton, word processor extraordinaire, to handle massive printing jobs for the study course. This text project, therefore, has benefited from the CIRM study course support through TTU. Two graduate assistants, Doyle Hunter and John A. Welch, helped with tables and graphics, and their support is much appreciated.

The Team

As Sir Isaac Newton said, "If I have seen further, it is by standing on the shoulders of giants." This Irwin text on integrative facilities management would not have seen the light of day without the serious and sustained efforts of its product development team.

The series editor, Dr. Thomas E. Vollmann, has provided rich precedent for quality research through his top-notch publications on production and operations management. He reviewed the manuscript of this book and guided its preparation, while positioning it in the overall CIRM program and sharing what other series authors were doing. Jeffrey A. Krames, the editor-in-chief for Irwin Professional Publishing, kept me moving.

"No man is an Island" remains true in our household, and so I owe a great debt to my wife and editor, Margaret. A creative writer, she struggled through the hundreds of pages of manuscript, improving its readability and eliminating repetition, to which I am prone.

Deep gratitude must be expressed to the late Donald W. Fogarty, Fellow of APICS, author of well-respected texts in the field and truly an inspiration to all who learned a great deal from him about the business and the personal side of being a dedicated APICS member. Don was a gentle but insistent impetus to my own efforts. APICS and Irwin have supported this effort financially, but it could never have been achieved without the experiences of decades of industrial involvement. For the last 18 years, I have been an active APICS member, and to the many APICS colleagues in both practitioner and academic ranks, I can only say thank you!

As to the flaws, they weren't anyone's fault but mine.

John M. Burnham

CONTENTS

SECTION 2 ENABLING

SECTION I

STRATEGY AND REPRESENTATION

CHAPTER 1

FACILITIES MANAGEMENT AND STRATEGY

When a company decides it needs a new plant, a new product, or new space, a group must be chosen to integrate all the various functional specialty areas into a coherent whole. Specialists are chosen by department heads as suitable people to represent their areas in group decision making and planning.

Assembling a group of people is a necessary first step, but it does not create a team. In a large company, those assigned are unlikely to know each other, and the project manager is likely to have come from another facility. Production, product design, finance and accounting, marketing, maintenance, process and manufacturing engineering, logistics, and human resources—all must be part of the facilities development group.

The group may well meet in a room on the shop floor so that operators can become involved in the many details that will make the undertaking a success. To become a team, group members must not only become acquainted with one another, but must understand what each person does, and how the company's strategy relates to the project. The CIRM book series should be available from the beginning of the team's development, although it might not be used seriously for some time—until need provides the motivation.

A STATEMENT OF STRATEGY

The aggregation of any company's physical facilities expresses its *strategy* and *purposes*. The term *integrative facilities management* (IFM) has been chosen to explain both this strategic function and the facilities manager's more traditional ongoing tasks of definition, planning, execution, and operations support.

The new strategic IFM responsibility is both challenging and refreshing in its import, providing a significantly different view of this world than readers may have been visualizing. It has become necessary to accept greater interdependence among all parts of a company and its external systems. It is a world with a much decreased timetable in which to make decisions and to develop and implement plans. It is a world that respects excellence and punishes mediocrity, where quality and responsiveness dominate cost and high-volume manufacturing. And it is a changing and an improving world as new and better products flow into the marketplace at an ever increasing rate, requiring strategic IFM decisions to support their effective introduction.

The Kaizen, or continuous improvement (CI), learning and betterment activities within manufacturing provide feedback affecting product development and other activities. An increase in the percentage of value-added activities in the plant will contribute to shorter lead times, better quality, more flexibility, and lower cost. And with these improvements, customers will permanently alter their expectations!

A world of change requires managers of change. IFM support to a dynamic manufacturing system requires continuous modification—an ever present responsibility. Further, it requires performance planning and performance measurement systems that match the survival needs of the organization.

INTEGRATIVE FACILITIES MANAGEMENT—THE BOOK

In addition to examining the vital strategic elements of IFM, this text will help develop your understanding of the concepts and the essential details associated with "the inside job": building, modifying, or adapting a manufacturing facility and managing it so that its assigned products will be made and shipped. This includes actively participating in the product and production design process. IFM must help with the facilities implications of the manufacturing concept and participate in selecting, laying out, and installing (or modifying and relocating) equipment and tooling to match the needs of the product. IFM is important in choosing, training, and supporting the production team that will actually make the product.

For all of these reasons, there will need to be a closer coupling of the strategic aspects of management and the associated manufacturing processes. Each reader is challenged to seek analogies in his or her own business and look for ways to keep industrial and office facilities in line with the corporate competitive strategy. Strategic examples include location choices, site layout and exterior appearance, and the suitability of office, public, and work spaces for both function and supportive "climate."

SITE DEVELOPMENT TEAMS

IFM's integrative focus treats the strategic, developmental, and operational issues affecting the company's physical assets used to carry out its manufacturing and marketing strategies. These issues are not neatly separable and are best managed by multilevel, multifunctional teams that span multiple areas, with shifting composition and emphasis to match evolving needs.

The term *site development (SD) team* is used to signify both the team and its defined mission. And while, significant gains in efficiency have been achieved through computer support—as shown in later chapters—most IFM accomplishments continue to be through *human* efforts, in teams. The initial team begins with a number of members from outside manufacturing who strive to represent the site viewpoint strategically. As the program develops, the SD team can be augmented by legal, financial, and real estate specialists and more process and project engineers, when specific greenfield or large grayfield programs are being contemplated. This larger team carries out the preliminary and detailed planning that enables the acquisition, construction, and start-up phases consistent with the mission. The shift toward operations brings with it a shift in emphasis toward support. The site support group members include safety, environmental, plant, maintenance, human resources, manufacturing, and industrial engineering personnel—and as with the other SD phases, this team will be augmented to fulfill its operational role.

DEFINING IFM

Integrative facilities management needs to be defined in its modern, multilevel context. In the narrowest sense, IFM installs and maintains the physical plant and its surroundings to make them consistently operable, secure, safe, and in compliance with all relevant regulations. The Library of Congress takes a broader look: IFM is "the practice of coordinating the physical workplace with the people and work of the organization; it integrates the principles of business administration, architecture, and the behavioral and engineering sciences."

The CIRM viewpoint is that IFM manages the installation and maintenance of process and support equipment at a site designed to fulfill the needs of office and production personnel. In turn, manufacturing must satisfy customer demands for the products assigned. This tactical level seeks to match needs with available resources. Location, scale, and other critical factors are givens to the solution framework.

Viewed as part of a dynamic future industrial economy, IFM's responsibilities include three distinct roles. At a strategic level, IFM is *representing* the IFM viewpoint realistically, participating as relevant strategic issues are analyzed. At a tactical level, IFM is *enabling* the location choices and site development, matching logistical needs, budget, and other resources constraints. Operationally, IFM is *supporting* the ongoing production, distribution, and improvement activities at the site, all these as an integral part of the manufacturing system. These three sets of relationships are outlined in Table 1–1.

Truly an integrative process rather than a functional activity, IFM works toward *decisions* based on inputs from the rest of the SD team *at the level being addressed* (representing: long horizon; enabling: intermediate horizon; or supporting: short horizon) and in the process generates the constraints—or the appropriate resources—that bound the lower-level solutions.

These bounds can themselves be the subject of high-level consideration and adjustment to reflect any strategic revisions. The materials-oriented reader might think of closed-loop Material Requirements Planning (MRP), while the quality-conscious reader might think of Joseph Juran's "fitness for use" spiral. Of course, the strategist will accept that facilities, being the expression of strategy, will be both decision and constraint as a part of the strategy process (Mintzberg and Quinn).

The columns in Table 1–1 reflect IFM activities and influences or actions from other sources: external (outside the company or, at the very least, outside the facility), manufacturing (principal customer of the IFM team and its activities), and internal (within the company, but not generally located within either manufacturing or the IFM team).

Some of the figure's terms—shorthand for significant concerns, concepts, or activities—may be unfamiliar to you. A brief walk-through of IFM involvement will set the stage for the details presented in later chapters.

TABLE 1–1
Multiple Roles for IFM Achieved Through Multifunctional, Multilevel Teams

Specific tasks in:	*External*	*Internal*	*Manufacturing*
Representing (Strategy) Capacity evaluation Location evaluation Logistic system stability Scale and scope Size and output est. Cost estimates Continuity plans	Capital Competitiveness Regulators Communities Environment By-products Charter	Capacity strategy Process technology Logistics strategy Networks of plants, suppliers, distributing facilities Output Focus Criteria	Key tasks Facilities strategy Scale, scope Manufacturing Concepts Eliminating waste Mission
Enabling (Tactics) Size, arrangement Process equipment and organization Structure and layout design Contract plans and specifications Project management	Location Logistics Individual site strategy Site development (facility, plant) Flexibility	Logistics Product lines Manufacturing concept HRD Coordination with other manufacturing systems	Suppliers Customers Layout Modularity Flexibility Phases
Supporting (Operations) Continuous improvement Proactive maintenance Plant data bases safety maintenance plans, layout	Dynamics Contingencies Personnel safety Regulation Community Insurance	Dynamics Contingencies Plant TQM,* TPM,* SGIA* Site development team	Dynamics Contingencies Continual improvement Quality Personnel development

*Total Quality Management
*Total Productive Maintenance
*Small Group Improvement Activities

DRIVERS FOR INTEGRATIVE FACILITIES MANAGEMENT

Table 1–1 presents *representing, enabling,* and *support* dimensions as those to be met by IFM. At the long-horizon level, action plans are developed cooperatively, with IFM as a proactive representative in, for example, strategic discussions about capacity, scale and scope, and location. Other plans are driven by uncontrollable external forces, such as government regulations, competitive posture, and the economy. Here,

IFM will seek to anticipate and plan for various alternative responses to these uncontrollable forces. (This kind of contingency planning is a major subject of Chapter 4.) Both the proactive influencing and the contingency planning of possible responses are strategic, representing roles for IFM. To understand just how these external and internal strategic forces serve to guide IFM, the broad company-wide strategic assessment process needs to be examined.

Corporate business strategy reaches far beyond the facility into the external world of investors, regulators, competitors, and the consuming public. Other influences are regional construction costs, operating economics, and access to resources and to customers. Product, process, and facility design choices can impact these factors. And where strategy dictates, the company, accepting the accompanying risk, may elect to disregard some aspects altogether.

Elements of Business Strategy

Companies increasingly use a method known as *competitive benchmarking* to assess their own capabilities against those of competitors. Strengths, weaknesses, opportunities, and threats (SWOT) of each of these capabilities are carefully analyzed in conjunction with such business objectives as return on investment, desired growth rate in market share, technological leadership, and new product introduction goals.

The outcome is a business strategy statement, typically including a portfolio of products and markets the enterprise wants to compete in and the kinds of assets, both physical (facilities) and financial, the enterprise wants to acquire. It is also likely to set goals for growth in market shares and capacity, new-product development, and introduction schedules.

The *markets and products strategy* must define choices of consumer targets, product attributes, customer value assessments, and competitive products and services. Pricing, the choices and characteristics of products to develop, their quality and durability, and the frequency of product modification or termination represent a set of "specifications" for how the product relates to the consumer. Competitive considerations will address the preferred manufacturing and distribution locations for "presence" and for prompt service. Such choices consider other substitution products and the key success factors for the company. These decisions drive IFM analyses on scale and scope, location, product manufacturing assignments and frequency of changes, and the inherent

TABLE 1–2
External and Internal IFM Strategy

Role	*External*	*Internal*
Strategic (Representing)	Customers Location Capacity Quality	Capacity Network Existing vs. New
	Competition Costs Technology	Location Logistics
	Environment Regulation Skills	Resources Technology Funds

flexibility the facility requires. Table 1–2 provides some details of this strategic level.

The *manufacturing strategy* will give focus to key tasks and technical competences of the firm. Through delivered products, manufacturing makes the marketing efforts a reality. Quality, product availability, flexibility, responsiveness, and cost—these choices drive IFM, principally in support of process design and manufacturing itself. Equipment choices, organization and installation, equipment maintenance and parts management, structure and layout that fit the manufacturing concept, overall plant efficiency, and lowest life cycle cost, and operations that meet zoning, safety, health, and environmental regulation—all are determined in conjunction with facilities plans derived from manufacturing strategy.

The *technology, research, and development strategies* of the firm consider products and manufacturing processes balanced for optimal contribution toward strategic objectives. Product and process design, introduction of new technology, materials changes, ongoing simplification and value engineering, and major product and quality improvement efforts are factors influenced by technology/engineering. These choices all drive facilities decisions about structure, "clean rooms," emergency washdown areas, layout, materials handling, provisions for utilities and energy use, and, quite frequently, the need for by-product analysis and disposition. Some details of IFM strategic considerations relating to manufacturing are shown in Table 1–3.

Financial strategies provide limits (availability, related to return targets) and means (alternative ways to finance when strategy requires it) as

TABLE 1–3
Manufacturing and IFM Strategy

Role	*Manufacturing*
Strategic (Representing)	Key tasks Responsiveness Quality Variety Costs Competences Technologies Experience Mission Competiveness Time-to-market (production)

TABLE 1–4
Finance and IFM Strategy

Role	*Financial Aspects*
Strategic (Representing)	Investment criteria Life cycle costing Resource limits Alternative-facility combinations

bases for analyzing investment proposals. Functional strategies will interact with the financial ones since long-term viability and profitability are mandated by the owners of the firm. In a nutshell, the investment of capital in fixed assets must pay off. Capital availability and rate-of-return targets are parameters that drive IFM decisions about location, initial scale, structure, and internal organization (equipment, offices, materials handling). Some of these IFM relationships are suggested in Table 1–4.

Logistics Strategies

Factors influenced by logistics include location, preferred product volumes from each site, ability to handle one or several transportation modes, and the number and size of manufacturing and distribution locations and their positioning relative to inbound and outbound transportation (railroad spur, truck road, airport runway). These factors drive

TABLE 1–5
Logistics and IFM Strategy

Role	*Logistics*
Strategic (Representing)	Distribution strategy Product volumes Variety Supplier networks Materials flow concepts Manufacturing Concepts Scale Scope

facilities decisions on size of site, overall building organization and expansion phases, number and nature of wall penetrations for point-of-use delivery, and materials handling needs. Logistics is a continuing factor at each of the three IFM role levels, but it has vastly differing concerns, impacts, and details at each level. The strategic (representing) issues are addressed in Table 1–5.

Regulation

Environmental protection, health and safety, state and local labor rules, zoning, and by-product disposal all have regulatory aspects driving IFM decisions about office and manufacturing space layout, the guarding of manufacturing and materials handling equipment, heating, cooling, and ventilation needs, filtering, clean rooms, fire and emergency equipment and escape routes, emergency procedures, washdown showers, and sources and disposition of hazardous gases or wastes. Some of these concerns are reflected in Table 1–6.

Stakeholders

Customers, employees, community, and owners are strong influences on IFM, especially as the facility is the company's face to the public—the expression of corporate strategy. The attractiveness of its site layout, upkeep, parking-space management, lighting, security and safety, and ease of access are important concerns. Transport truck routes, frequencies, and appearance and employee vehicle movement can be a source of either great concern or stakeholder satisfaction.

TABLE 1–6
Regulation and IFM Strategy

Role	*Regulation*
Strategy (Representing)	EPA Water Pollution Air Dispersion Land Disposal OSHA Safety Environment Noise Emergency Access (ADA) Company Policies Labor Law Zoning

Management

Company values—"who we are; what we think is right"—drive strategy. The impact of corporate-level policy on plant siting is usually significant, especially for high-tech companies. Where specially skilled employees are a key to viability, they must be recruited and retained. In addition, both location and scale can be governed by such policy guidelines.

Human Resources

Human resources management (HRM) is concerned with overall management personnel policies and practices, safety and emergency provisions, industrial and bargaining unit relations, and the workforce numbers and skills to support manufacturing. These concerns drive IFM decisions about office and manufacturing space allocations and finishing, meeting and training rooms, communication systems (video monitors, information boards, audio or music broadcasts at the site), and personal lockers and recreation facilities.

Perhaps the most significant challenge raised by HRM is the level of skills required to operate and maintain the plant, and whether such skills are available in the prospective community. This will affect the potential scale of operations, the degree of vertical integration, and/or the magnitude of the training effort to support plant operations. For many industries, the community environment—cultural, educational, recreational, and spiritual—may be the dominant factor in choosing a location.

TABLE 1–7
IFM Roles and Interactions Summary

Role	*External*	*Internal*	*Manufacturing*
Strategic (Representing)	Customers Competition Environment	Capacity Location Resources	Key tasks Competences Mission
Tactical (Enabling)	Markets Site development Flexibility	Contingencies Modularity Phasing	Scope Products Responsiveness Manufacturing concept
Operational (Supporting)	Dynamics Environment Constituencies	Adaptability Safety, by-products LLCC (lowest life cycle cost)	Continuous improvement Capability Site support

This array of external and internal factors creates specific or general guidance for the various levels of IFM strategic and tactical plans, implementation, and facilities operations (Table 1–7). It cannot be ignored.

CRITICAL FACTORS FOR OPERATIONAL SUCCESS

Whatever the successes in planning, IFM will be evaluated along the following significant results-oriented dimensions, all supporting the manufacturing customer over the economic life of the facility.

Construction/Completion

For both greenfield and grayfield projects, meeting the construction timetable requires civil, manufacturing, and industrial engineering disciplines, *integratively* applied and in combination with purchasing (for suppliers), logistics (for both supply and distribution), production planning (for feasibility and practicality of workload), and manufacturing (for consistent execution of 100 percent quality work). The key criteria are adequacy of the facility (roads, docks, point-of-use access), capability (equipment and tooling that can make product 100 percent acceptable), capacity (equipment, materials handling, passageways, layout that can easily accept the volumes required), and maintenance (to support the manufacturing equipment, processes, and people).

Production Planning Support

Managing work completion to meet the timetable (construction, commissioning, modifications, maintenance), staying within budgets for all areas of responsibility, and maximizing equipment availability (uptime) against the production schedule are mandatory. Construction timetables must fit with production. For example, by the second day after turnover at Blytheville, Arkansas, the Nucor heavy shape mill was providing marketable finished product—this in just over a year after the start of construction.

Positive Regulatory and Community Relations

Maintaining and enhancing the facility's positive reputation with government and community representatives require a combination of *awareness* (currency on federal and state regulations, local rules, zoning, intensity of use), *reflective thinking* (ways to anticipate and constructively resolve issues raised by the community and by regulatory changes), *analysis* (changes in volumes, processes, or products; vehicle and human egress; hazardous materials), *record keeping and database development,* and proactive *improvement programs* (health, safety, environmental work standards, and by-product utilization and recycling).

Within the IFM activity there must be a knowledge of regulatory requirements and intent. Local ordinances or fees for vehicle passage, waste disposal, or utilities; air and water quality standards; and requirements for packaging or truck enclosures can be either anticipated or reacted to. Participation with regulatory technical groups examining hazardous materials handling and processing is a positive response to proposed standards. Attempting to "water down" regulations to reduce their impact is not.

Support of Manufacturing Dynamics

A principal feature of today's manufacturing is its dynamics, in response to both external forces and internal improvement efforts. For example, this is reflected in the need for proactive emergency/disaster planning on the shop floor. The many changes taking place almost daily can result in obsolete plans for escape, countermeasures, and protective operations should disaster occur. New layouts, modified equipment, closed-circuit ventilation for hazardous vapors, and closed-loop recycling for dangerous fluids all require prompt updating and dissemination of plans.

Equivalent issues arise when addressing risk (insurance) planning. Mission changes, new or terminated production assignments, equipment additions, modifications, and deletions and training require risk reviews for both property and personnel. Policy provisions may stipulate OSHA compliance as a condition of insurance coverage. Steel mills have both safety and environmental hazards that must be dealt with effectively through both design and operating safeguards.

Responsiveness

The various aspects of flexibility and their consequences mandate an integrative approach to all of the engineering functions, including IFM. Some process plants combine all but design engineering into a team of industrial, manufacturing/tool engineering, and facilities engineering. This shortens response times and reduces after-the-fact "debugging."

IFM Managers

Each IFM participant is inextricably bound up with the company's physical facilities. Being very long term, facilities decisions have a lasting effect on the company's competitiveness. After original development, major changes to the building or site layout are infrequent. Even less often does a company change the location of a plant.

So there is a need, at the original facility design stage, for truly integrative analysis with almost no restriction on the range for decision making. The facility is a very long term part of the logistics as well as the production system and must respond to changes in technology, markets, products, distribution, human resources, regulation, and the environment. And changes require both design and operations support at the site.

Change

In addition to important traditional challenges, IFM must meet the demands of the customer, for whom both the external and internal environments of 21st-century manufacturing are increasingly dynamic. Manufacturing and the IFM "supplier" must face global competition, time compression and short product life cycles, impacts of technology, and new ways of bringing products to market. Manufacturing must also accommodate the need for continuous change. Therefore, facilities managers face a new challenge: to develop and maintain facilities with decades-long economic lives, when the products to be made in the facility sometimes have life cycles that are as short as a few months!

In global competition, the manufacturing customer is facing radical changes in requirements. To preserve structural, environmental, and utilities' effectiveness over the life of the plant, IFM must develop and sustain a high degree of flexibility in facilities' design and execution. The financial customer will remain concerned over project cost, maintenance, and modification budgets. Manufacturing operations strive for continuous improvement, which affects IFM equipment and safety record keeping, maintenance practices, personnel development, engineering data on layout, utilities' support, budget, and staff size. To provide more direct and specialized support, individual IFM and maintenance staff may be co-located at various factory units.

Shortened Product Life

The Hewlett-Packard Company has noted that over half of its product line is less than five years old. Cal-Comp has backed away from mass assembly techniques and formal materials planning systems in favor of product simplification and low-tech, flexible manufacturing. In the traditionally stable textile business, Collins & Aikman Company's specialty fabrics business has moved from a principal mix of lingerie and action wear fabrics to a highly sophisticated auto upholstery material—this in considerably less than five years in a 20-year-old commodities–yard goods plant.

These realities require responses in both design and adaptiveness of the manufacturing facility. Plants must not only house the original production equipment and deliver products, but do so for a whole succession of products over the life of the facility.

People

There are other challenges. The combination of just-in-time (JIT) manufacturing and total quality management (TQM) has caused a number of office personnel to be co-located with the production facilities they serve. Engineers and production planners, maintenance specialists and computer programmers, plant unit managers and supervisors are found in increasing numbers right in the middle of the production space. This means providing work and meeting spaces where none have traditionally been needed. And the existence of "offices" within production areas again emphasizes the integrative nature of facilities management.

Recognition that the company can benefit from grass roots input by operating employees has opened up yet another IFM challenge. Small-group improvement teams work for continuous improvement in both production and support. This requires meeting rooms, training rooms, and on-the-spot discussion areas where a group can get together to solve a problem.

Other clear trends address TQM, total productive maintenance (TPM), and self-directed work teams (SDWT). TQM seeks a high-quality product and calls for data gathering, experiments, and problem solving to take place throughout the plant. TPM seeks the same sort of result in the equipment and space maintenance areas. And SDWT has the goal of task setting, planning, executing, and controlling substantially all aspects of production without supervision.

Each of these efforts requires a supportive working environment and facilities. Office design has for at least the last two decades recognized the dynamics of organizational restructuring and the need to provide a supportive environment for building occupants. Many office design approaches must be applied to the production area: personnel access routes, safety and security, ventilation, lighting, construction standards, and finishing detail.

Simultaneous Engineering and Teams

Simultaneous (or concurrent) engineering (SE) is increasingly being applied to product/process development. To make the design process faster and more effective, simultaneous engineering assembles a team that includes marketing (and sometimes even the customer), product development, product and production design and engineering, manufacturing, facilities management (IFM), and, often, existing suppliers and purchasing representatives. For instance, at Collins & Aikman's Specialty Fabrics Division the product development group is housed in the plant and makes trial (sample) runs of new fabrics on regular production equipment. The implications for IFM are clear: Not only does IFM have to be a multifaceted team effort, but it must be synchronized with simultaneous engineering in product/process development efforts.

Simultaneous engineering can serve to enhance IFM enabling and supporting activities. Much of the missed communications and lead time consumed in the traditional development process for the individual facility can be "collapsed" into the work of a site planning and implementation team. SE/SD will reduce the functional barriers among the many communities that must be involved and will stimulate communications systems that will keep all relevant parties adequately informed. Figure 1–1 shows the "new design loop" that is involved.

IFM's challenges—most notably those of capturing and understanding the many points of view, criteria for success, and problem-solving approaches, and reflecting these concerns at many hierarchical levels and in many contexts—can be addressed using the SE/SD approach. The site development team is expected to accomplish the effective integration of the many relevant viewpoints needed for a robust, flexible facility.

TRENDS IN BUSINESS AFFECTING FACILITY DEVELOPMENT AND MANAGEMENT

External and internal factors that impact on the company also affect its facilities strategies and plant design and operations. These include the drivers mentioned previously and are generalizable into the following key trends.

Total Quality Managment

Major companies worldwide have undertaken the JIT journey, most with considerable success. There is increasing pressure to improve quality and reduce lead times, thus improving flexibility, producing better and lower-cost products, and being much more responsive to the customer.

Some people prefer to combine JIT and quality mandates into JIT/TQM—or, simply, total quality (TQ). The issues with respect to quality are both broad and specific. IFM involvement occurs at all levels. Air quality and environmental conditioning can drastically affect product quality, as well as operator safety and health. And equipment modification to improve the quality capability of a manufacturing process often will involve IFM. Teams of operating people address, analyze, and solve problems. There is a strong trend toward operator-centered quality control, again shifting the emphasis to more integrative thinking and flexibility choices. Proactive IFM participation in both the design and

FIGURE 1–1
New Design Loop

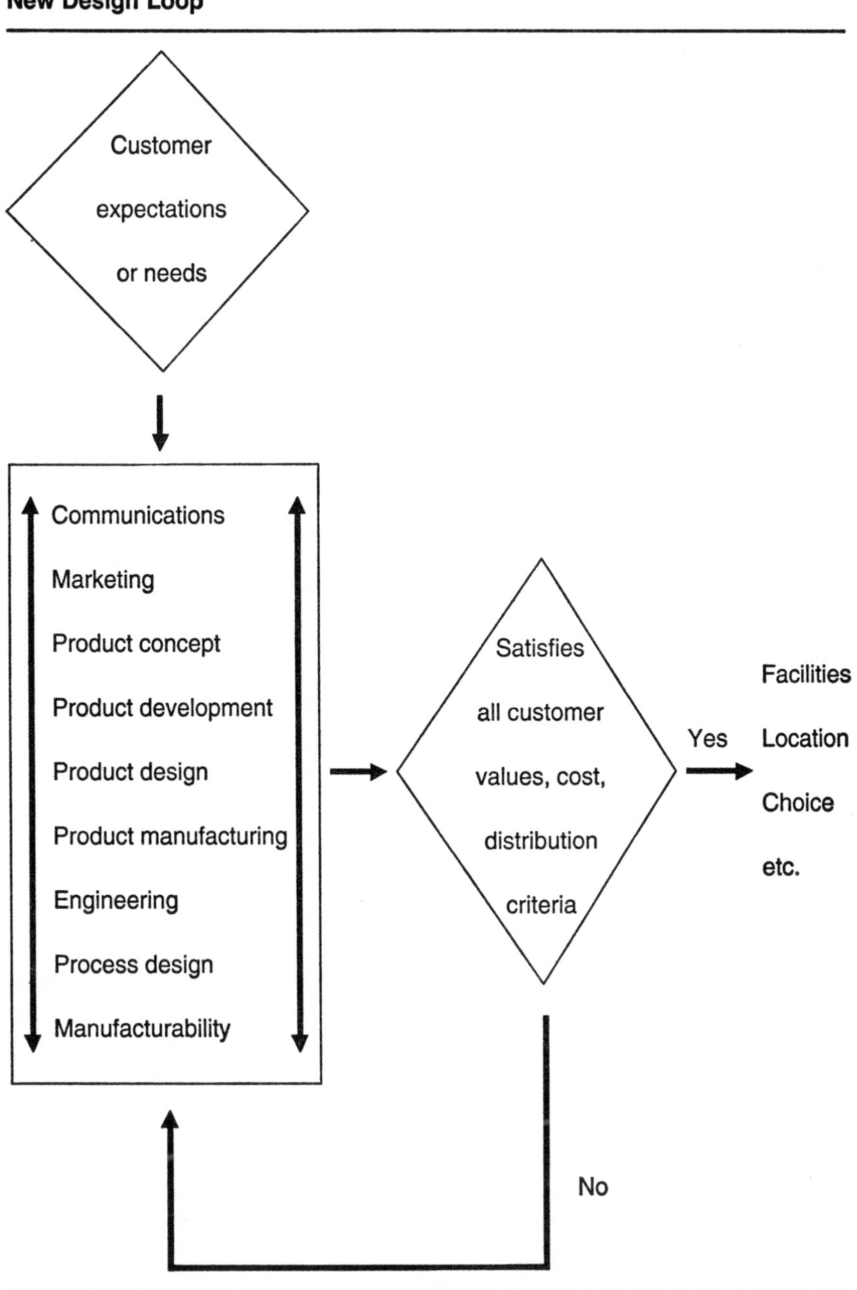

operating phases of the plant life is needed to achieve optimal process and product quality.

Total Productive Maintenance

A parallel to the TQM trend is that of teamwork to achieve sustainable and optimal equipment performance through operator-centered maintenance. The same operator-team approach studying and solving maintenance (downtime) problems has been very effective. This trend affects IFM both in space allocation and in maintenance technician management and staffing. The plant is planned for the best relationships between outputs, space, manufacturing costs, human resources needs and capabilities, and external (societal and environmental) considerations.

Changing Organizational Form

The reduction in layers of management and the increasing need for teams rather than functional task assignments have resulted in much more fluid organizations, in offices as well as on the factory floor, and in flexible facilities for team discussions, meetings, training, and so forth.

The integration of office/support staff with the production workforce is another aspect of the changing organizational form. Co-location of engineers, materials buyer/planners, cost accountants, and unit managers to the factory area where their product is made allows quick coordination, consensus, team building, better problem identification and solution, and more opportunities for involvement and improvement.

Cellular Manufacturing

Perhaps nothing has had such a profound effect on the modern manufacturing plant as the rediscovery of the many benefits of cellular manufacturing. "Cells" of dissimilar equipment are organized to enable smooth, flexible manufacture of a family of parts or products. This fits beautifully into the concepts of work teams, self-management, small group improvement activities, simplification of scheduling practice, reduced throughput time, and flexibility at the work cell. Quality is improved in the cell environment because of quick feedback and inherent teamwork. Cells also fit well with a related concept, group technology (GT), seeking the fundamental commonalities among parts or assemblies so that smaller, more flexible machine groups can be managed together. And it reinforces the power of Professor Wickham Skinner's focus factory concept, since each cell, or GT group, is really a plant-within-a-plant and can be managed for its internal similarities.

New Workplace Designs

New workplace design concepts and details facilitate team meeting spaces, the co-location of support team members with the operating personnel, the physical reorganization of the plant layout for cellular manufacturing based on GT analysis, and the need to accommodate change flexibly and effectively. All these call for a proactive IFM stance and considerably more contingency planning so that the operating facility remains efficient in the face of changes.

Life Cycle Costing

Historically, facilities and equipment were developed to meet long-range planning objectives based on the ability to work with 10-to-15-year product life cycles. These product cycles, extended through modifications by marketing and product design (cosmetic, or features, engineering) meant that capital justifications were direct and results measurable through product success and plant operating costs. With experience, especially in the decades following World War II, sales and cost estimating became increasingly reliable. In more recent years, however, the combination of oil shocks, increased global competition, and constantly improving products has driven the product life cycle downward. For the typical facility, neither mission nor products are predictable over the building's life. So shorter product life cycles have led to the need to relate facility costs to a succession of products.

The logic is that rather than justifying the facility once and for all, it becomes vital to associate specific products with the equipment, services, and personnel needed to create and support them over each assigned product's life. Lump-sum calculations for the plant thus give way to the need for clearly defined product-related costs, these to be compared as the basis for proceeding with revenue projections. The costing basis has also changed, from absorption by department toward activities required to support production—the activity-based costing (ABC) system from Johnson & Kaplan's *Relevance Lost: The Rise and*

Fall of Management Accounting. And the most current emphasis is toward activity-based *management,* not tied exclusively to costing.

Environmental Activism

Environmental activism is an extension of the traditional IFM responsibility for the company's face to the community. "Love Canal," acid rain, and the issues of nuclear waste disposal are indicators of the extent and seriousness of this concern. IFM professionals must engage in ongoing systems analysis for *all* inputs and outputs from the plant manufacturing system and actively monitor changes in regulations, new health problems, and community concerns affecting the plant or its employees.

The overall objective is safe, uneventful management and disposition of all by-products, including a reduction in the quantity of waste, process changes to eliminate hazardous effluents, and even conversion to create a marketable by-product out of process waste. (Chapter 5 addresses a proactive IFM analysis of waste.)

IFM representation in the strategic councils is not solely that of knowing what to expect, but putting forward initiatives to benefit the entire logistics system. For example, decisions regarding scale, scope, and location have profound effects on construction, operations, and transportation costs, as well as on being customer responsive.

IFM roles are still unfolding, but some directions are clear. Flexibility and short product life cycles, continuous improvement and emphasis on added value, policy deployment and simultaneous product-process development—all these mean that facilities must accommodate continuous change. Operator-centered changeover requires space for tools and fixturing. Point-of-use delivery and storage leads to changing space allocations, with more curtain walls and side receiving aprons. Materials handling systems cannot freeze materials flow. Lighting, ventilation, and consumables piping must have flexibility near the points of use. Drains and pressure piping must be carefully engineered to minimize the cost of future changes. Equipment installation should be safe, but as temporary as the specifics will permit.

IFM Operational Activities

As the focus of IFM turns to a support role, such questions as scale of operations, procedures, equipment and process steps, tooling procurement, materials planning, supplier selection, and, finally, operations planning have already been settled. In coordination with marketing (the customer), the production planning and control system can begin functioning to schedule the product(s) for completion and shipment.

Discipline is required for a system that maintains equipment capabilities, exercises operating control of equipment so that quality is maintained in execution, and manages the whole facility in a safe, environmentally sound manner—in addition to handling production planning and control, physical distribution management, and customer service. The right product must get to the right place at the right time and in the precise quantity needed for customer satisfaction. Most of these systems will operate in near real time and require considerable attention to do it right the first—and *only*—time!

PHYSICAL ASSETS: THE EXPRESSION OF STRATEGY

Physical facilities—whether terminals, offices, manufacturing plants, sales offices, or distribution warehouses—collectively embody the company's competitive strategies, its value system, and how it seeks to relate with employees, customers, suppliers, and the public at large. A key weapon in seeking market advantage is the analysis of the pattern of facilities decisions made by competitors, which can lead to a fair approximation of their strategy!

Today, corporate headquarters and company sales and administrative offices in general have moved toward a silent communication about "who we are." Architectural and engineering design firms have successfully applied the tools of computer-aided engineering (CAE), and computer-aided design/computer-aided manufacturing (CAD/CAM), to commercial buildings and office spaces. The results since at least the early 1980s of post-occupancy surveys have been consolidated by computer to suggest design improvements and ways to improve the customer match during the inevitable changes in a building's life. This practice also must increasingly be applied to the industrial facility. The frequency and magnitude of change—50 percent of offices go through significant modification each *year*—may surprise manufacturing people, even though they are well aware that a company's commitment to JIT manufacturing and TQM brings about the need for changes to the industrial plant or warehouse on almost a continuous basis.

Other forces require change in the industrial facility—for example: product mix, demand growth or decline, new or discontinued products, value-based redesign, new (sometimes threatening) technologies, and competitor actions. Although some creative ideas have been implemented with great success, they seem to be the exception rather than the rule. The evidence is incomplete, but it appears that computer technology has been applied to industrial facility design less often than to office spaces—the exceptions being group technology coding as a basis for manufacturing cell development, and space layout programs, such as CRAFT and CORDELAP, which optimize space assignment based on transaction frequency and travel distance. Just as office design addresses the needs of people, industrial facilities emphasize the product to be manufactured rather than the equipment, layout, and people who carry out the manufacturing task. The opportunities for improvement are significant.

CHAPTER 2

CONFLICT AND RESOLUTION: THE SITE DEVELOPMENT TEAM

As the group begins to settle into what must be done, the many conflicting perspectives generate lively discussions, which help team members become further acquainted and expose them to work and ideas beyond their own experiences. Skillfully managed, these discussions can lead to shared understanding of major and minor tradeoffs and how these will affect the overall project.

Almost from the beginning, there will be a need for specialists to fill assignments in other areas, such as environmental sensitivity and analytic capability, but there will not be sufficient workload to justify added persons. CIRM book study can help all of the group to become increasingly aware of these and other dimensions, and allow the deferring of team additions until the requirement for regular support arises.

THE CONFLICT POTENTIAL

It is not surprising that in a 1991 survey of manufacturers in the Tennessee Valley the most important areas targeted for improvement included team building and communications. Cooperation and coordination among the interdependent entities involved in integrative facilities management offer the only effective way to create a good plant that provides customer satisfaction. Any *involved* group will certainly strive to do a good job, for the company and for the members' own interests, but conflicts will arise through problems of communication, discipline-acquired language (jargon), performance measures, time frames, and explicit (and overlapping) functional responsibilities and authority.

Table 2–1 is an expanded version of Figure 1–1, building on the materials covered in Chapter 1. It develops the various conflict issues facing any industrial group, and it is the principal justification for having this new CIRM program.

To illustrate the potential for conflict, consider the planning horizon and the time frame for decisions just among the IFM and the manufacturing process development (MPD) and operating production (MFG) personnel. These three specialties have many shared viewpoints and are grouped together within the manufacturing processes module of the CIRM program.

IFM has direct awareness of capacity and facilities strategy reaching years into the future. Once constructed, the facility has decades of economic viability, with lots of possible grayfield modifications ahead.

MPD is driven by a dynamic product-idea-to-development cycle with a much shorter time frame than IFM for the associated process design decisions and their implementation.

And MFG—while participating in both plant construction and layout and the process implementation through equipment and systems installation—deals with an even shorter horizon, perhaps just days.

The idea of a plan, then, has enormously different meanings to each functional participant. Consider how the term *flexibility* is interpreted and brought into practice by each group.

IFM thinks in terms of modular building elements, service spines, temporary shop floor systems, and planned expansion as a part of the building or its eventual site development. (See, for example, Hewlett-Packard's "campus" concept, detailed in Chapter 3).

MPD anticipates new products and variations and addresses flexibility through equipment and fixturing choices as well as manufacturing cell formation.

And MFG has fewer remaining degrees of freedom and becomes most dependent on the materials support infrastructure, on the one hand, and on the increasing flexibility of the *work force,* on the other.

Origins of Conflict

Why do conflicts exist? Different functions have responsibilities in the same areas—location, layout, equipment, materials handling, energy, and how to operate—but with different perspectives. Corporate staff

TABLE 2–1
The Multiple Roles for IFM

Specific Tasks	*External*	*Internal*	*Manufacturing*
Representing (Strategy) Capacity evaluation Location evaluation Logistic system stability Scale and scope Size and output estimates Cost estimates Continuity plans	Capital availability Competitive pressures Regulatory mandates Communities (jobs, taxes, appearance, noise pollution) Charter Company needs vs. economics, New opportunities	Capacity strategy Process technology Logistics strategy Networks of plants,suppliers, distributing facilities Output needs Focus Consensus	Key tasks Facilities strategy Scale, scope Manufacturing concepts Eliminating waste Mission
Enabling (Tactics) Size, arrangement Process equipment and organization Structure and layout design Contract plans and specifications Project management	Location choices Logistics design Individual site strategy Site development (facility, plant) Flexibility External appearance Regulatory community concerns	Logistics (site flow) Product lines Manufacturing concept Coordination with other manufacturing systems HRD	Suppliers Customers Layout Modularity Flexibility Phases (vs. do-it-all-at-once, minimize interruptions to manu-facturing)
Supporting (Operations) Continuous improvement Proactive maintenance Plant data bases Safety maintenance plans, layout Logistics support	Short product lifecycles Contingencies Personnel safety Regulation Community Insurance Materials and vehicle movement Grounds upkeep	New products, technologies Contingencies Plant costs, efficiency TQM, TPM, SGIA Site development team Plant maintenance Inventory man-agement (MRO)	Operations dynamics Continual improvement Quality Personnel development Housekeeping Predictable performance Proactive maintenance

(finance, logistics, technology, manufacturing, and marketing) also have an interest in many of the same areas, but with their own perspectives.

The SD Team (SDT) with its multihorizon internal-external viewpoint, also is sensitive to both financial (LLCC) and manufacturing (materials handling, layouts, and production systems) concerns. This eclectic point of view is not generally present among all the assigned staff. In the process of understanding the reasons for conflict, team members must work through their own functional predispositions and reach consensus on what is needed for company success, and what this means at the site.

IFM team members will never be totally immune to seeing situations in terms of their own disciplines, especially as greenfield or grayfield efforts move toward execution. For many organizations, this teamwork and consensus building demand a culture and climate not currently in existence.

Conflict Sources

Conflicts often originate in upper management, although the symptoms may occur below that level. Differing time frames for getting results, performance measures, and reward systems are major sources of such conflicts. Resistance to change inherently arises from not understanding *why* the change is needed, so there is a leadership challenge here as well. Obvious differences in the education and training of engineers, operations managers, and marketers can generate problems which only working together can ease. SD Team members will experience the same difficulties, will have to resolve jargon, performance criteria, and functional differences—before they can *become* a team.

Time and Cost-Based Product Design and Development

The translation of product satisfaction criteria into needed skills, materials, and equipment is the ultimate test of the integrated design and development team, which must generate agreed-upon goals and common performance criteria. The elements of technical satisfaction must be combined with time-to-market and cost criteria and with the current or potential manufacturing capabilities and capacities at the plant.

The natural planning horizons differentiating operations people from corporate personnel lead to a variety of potential conflicts that may reach back to the manufacturing and capacity strategy areas. Ultimately, it is up to production management to meet customer demand, but timing, justification, staffing, technology choices, and funds availability all may seem to stand in the way of progress at the plant. The site development team is sometimes caught in the middle of the cross-currents of differing horizon and performance perspectives.

Just as marketing people wish to have inventory ready to ship, so production people want to have, even with the inevitable late changes, enough capacity and production resources (materials and people) to meet schedules. These desires can also lead to alliances with marketing (in general, at least) and conflicts with accounting and distribution. And the ability to influence decision-making has considerable effect on site development. Nonmanufacturing space is part of structure cost, and inventory is still costing *somebody*.

Manufacturing Processes

Within the three elements of the CIRM manufacturing processes module, there are frequently differences in horizon, criteria, and viewpoint. The nature and scope of these relationships and differences are shown in Figure 2–1.

The internal factors are the interdependent peer activities of manufacturing process design and of the manufacturing operations themselves. This set of relationships is illustrated in Figure 2–2.

Detailed manufacturing process design and development depend on three important corporate influences: marketing through product development, engineering through R&D and technology, and overall manufacturing strategy through key tasks and technical competences. Manufacturing product/process design should be done by team members representing all of the strategic areas and other internal and external areas too. Financial analysts, process engineering consultants, potential equipment and materials suppliers, production (manufacturing) engineering experts, human resources professionals, and packaging and distribution specialists are certain to be involved. In Chapter 1, this approach was described as "simultaneous engineering," and it is a very effective organizational mode for product-process design and development. Some companies also solicit input directly from supplier specialists, field service personnel, and the customer.

The details of manufacturing operations are the synergistic result of both policy deployment and tactical considerations. The manufacturing strategy and the manufacturing concept drive both the general configu-

FIGURE 2–1
External Influences Involving IFM

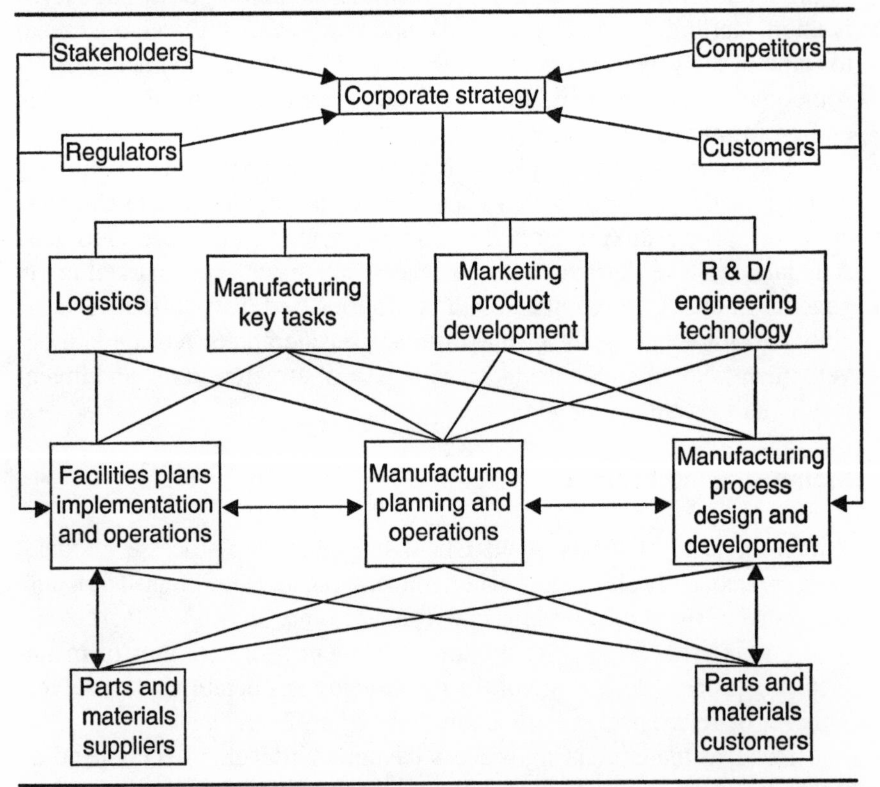

Source: John M. Burnham and R. "Nat" Natarajan, *Manufacturing Processes*, Student Guide (Falls Church, Va.: APICS, 1992). Reprinted with the permission of the publisher.

ration (layout) of the factory floor and its operating mode. Plant and workforce sizes are frequently set by policy, and the logistics function today is corporatewide and very influential.

Equipment choices (greenfield) and equipment modifications (grayfield) are led by *process design*, or by improvement ideas that may come from anywhere, including small group improvement activities in either quality or maintenance. And *facilities management* as a functional specialization starts with capacity strategy and location criteria and works through to facilities construction and support. These relationships are described in Chapter 3 and synthesized in Figure 2–3.

Beyond technical factors, manufacturing needs strong relationships with human resources in recruitment, compensation, and motivation.

FIGURE 2–2
Internal Influences Involving IFM

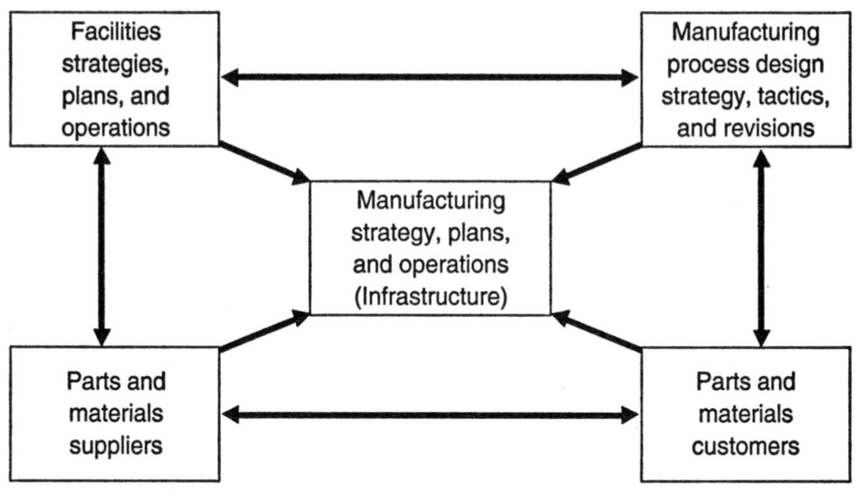

Source: Burnham and Natarajan, *Manufacturing Processes,* Student Guide (Falls Church, Va.: APICS). Reprinted with permission of the publisher.

Ongoing employee education efforts include training, development, retention, and promotion modes, principally related to cross-training, teams, and multifunctional tasks. Significant movement away from traditional plant structure and layout accompanies these manufacturing concept shifts, as engineers, buyer-planners, maintenance technicians, and supervisors are all being based on the factory floor. Training, SGIA, and problem-solving task teams also need meeting spaces near the work areas.

If Everybody's Responsible . . .

Plant layout, materials handling, manufacturing concept implementation, utilities design, and structure are all shared site development concerns.

The desire for manufacturing operations to have a cost-effective structure and materials flow system and an optimal set of manufacturing processes can lead to learning efforts perhaps even more intensive than those involving quality function deployment. The fundamental issue is that of different functional specialists having detailed, assigned responsibilities for certain aspects of the facility design and, quite often, different facility performance criteria by which their actions will be judged.

FIGURE 2–3
"Vertical" Relationships in Facility Planning

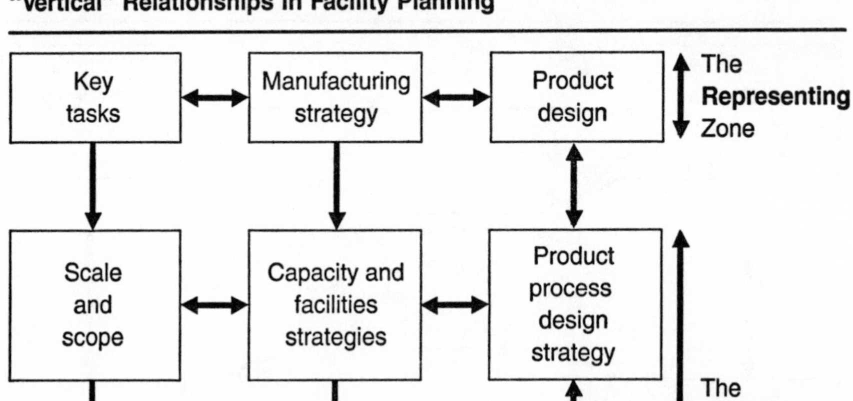

Source: Burnham and Natarajan, *Manufacturing Processes*, Student Guide (Falls Church, Va.: APICS). Reprinted with permission of the publisher.

Many different functional points of view about the in-common aspects of the facility include the process (and routing or sequence), the equipment, the way the equipment is organized (for operators, maintenance, and material flow), the size and structural arrangement of the building, the means of access, and the inherent environmental appropriateness and flexibility of the facility for accommodating those changes not anticipated when design was under way.

Conflicting Needs and Criteria

IFM and manufacturing operations (MFG) interactions include standardization, equipment capabilities, and labor skills requirements. When production and inventory management (P&IC) is considered, interactions broaden to include capacity, product mix, and the time required for changeover and maintenance.

Knowledge of equipment reliability traits can affect manufacturing engineering (MPD) in process design and in equipment specification. Manufacturing Engineering should know power, waste, and equipment expectations and communicate them as part of SD Team efforts.

MPD can affect staffing and skills required for operations (MFG) and maintenance (IFM). As noted, process choices can actually preclude some potential new manufacturing locations due to limited workforce availability or skills.

Human resources encompass labor skills demanded by process, equipment, and perhaps even layout. Skills being used should relate to new equipment/process planning and to cross-functional training, workforce flexibility. Related issues include standardization, labor relations policies, and safety, health, and environmental regulation.

Maintenance decisions can affect capability and thus quality of output (MFG). Maintenance plans, changeover capability, process reliability, and capacity are related. If equipment is down for repairs (scheduled or not), then MFG and P&IC need to know when it will be up again. MFG provides valuable feedback to MPD and IFM on equipment performance, capability, capacity, and support needs.

IFM decisions on heating, ventilation, and air conditioning and power, lighting, and utilities can affect MFG in terms of both employee morale and productivity.

IFM and MPD decisions that determine the facility and processes design will affect MFG. When the plant has been commissioned, MFG must get the needed output and be productive! This certainly will tend to motivate working with the SD team at various stages, even if this is inconvenient and time consuming.

SD teams must jointly consider the means for achieving quality levels: qualify function deployment (QFD)—volumes, mix contemplated *and range of variation*; lot sizes, proper design, installation of equipment, and tooling for shop use; preventive maintenance—training, safety enforcement, and tooling design, redesign for decreased changeover time, and quality improvement.

A Capital Budget: Built on What?

The traditional capital budgeting process for either greenfield or grayfield projects covers initial and time-phased expenditures, a timetable for startup, and the projected revenue stream (or contribution, taking operating costs into account).

The product-process development (MPD) derives from corporate marketers and technologists and then is turned over to development teams. The capacity and facilities strategies (IFM) provide guidance to individual plant charter and mission statements. And the manufacturing strategy and key tasks (MFG) guide those policy choices that conceptually configure the plant, all as shown earlier in Figure 2-2. It is no mean task to coordinate these three internal streams of activity. To the extent that teams are active at all three planning horizon levels, and draw on the viewpoints of these diverse specializations, then conflict gives way to consensus and "simultaneous engineering" can take place.

Note, however, the complexity of the issues to be addressed at various levels. At the representing level, *as new product ideas become development projects*: What will be the new product's life cycle? What is the total sales expectation during that life? How widely will the product be marketed? How widely and where will it best be manufactured? If at an existing site, what capacity and capability must be added? What will be the impact on existing products and customers? On staffing levels? Training? Current improvement plans?

At the enabling level, *as new market opportunities arise:* What should be the company's regional response? If new capacity is to be added, where? What site determinants are most significant, and how should a search be instituted? What about all factors of production in terms of availability and cost?

At the support level, as manufacturing strategy, technology, and key tasks suggest *different ways to configure production facilities:* What will the ideal new facility look like? If it is at an existing site, what portions can be retained? What effects will this have on production, staffing, expenses, suppliers and customers? How will this affect divisional performance?

Note that all of these questions are reasonable for the functions involved and require answers. What could be the consequences of developing the responses in a silo—that is, generating solutions to suit the functional area viewpoint, but not working closely *across* disciplines? Many companies can point to their own manufacturing systems as sad examples of this mistake.

The need for cooperation explains the real challenge of generating capital project proposals and budgets in today's dynamic manufacturing environment. Long-lead-time activities like plant construction must avoid agreement too early among MPD, IFM, and MFG, so that budget and product will match real needs. This cooperative effort includes contingency planning for the unknown, and both flexibility and modularity in equipment and facilities.

But how should cost projections be assigned against revenue-producing activities? Following activity-based costing, it has been suggested that the *fixed* plant costs include the site, civil engineering and construction costs, and the initial facility modules. Within the facility, there will be *process*-related costs recoverable through any product assignments using the process. Finally, initial *product* assignments should carry the costs of tooling, fixturing, and materials-handling equipment unique to the product but not to the general processes. The estimated direct costs of manufacturing are assigned to product.

Done with care, and with the company's gain in mind, the cost of provisions for future contingencies will be borne by *all* products, costs for families of products by the current and future members of that group, and those expenses for specific products directly by them, just as is done with labor, materials, and other production factors. Because of phased facility construction, expansion, and staffing, this costing process is consistent with Johnson and Kaplan's ABC concepts.

Although all of these costs are included in the capital justification budget, their breakdown into charge elements for cost accounting offers another opportunity for conflict that probably will not affect many in the plant itself but might well create friction elsewhere.

Life Cycle Planning and Costing

Tool/machine replacement, layout, and utilities are also concerns involving IFM, MPD, MFG, and finance.

But consider these issues: tooling and fixturing design may not match well with either financial or operating needs. A short product life cycle might suggest throwaway fixturing, while MPD design for a generic family of related parts/products might take more staff time and involve greater costs. The budgeting and costing for the first product might appear to penalize it, while being very much in the interest of the overall plant and product family.

Replacement issues follow the same pattern, needing a broader, more strategic approach than time usually seems to allow. These issues are somewhat similar: What is the financial, product, and manufacturing basis for the replacement decision? What is the lifetime requirement for this function? What technological and product life cycle issues must be considered? What quality capability now exists? What will exist in the future? Options that were not available at the time of original commissioning may now appear attractive.

Example

ANACOM, an Ohio-based maker of small machined parts, developed a realistic approach to its equipment choices: opt for the best technology available that *reduces the break-even volume*! The company made great strides with setup reduction to accompany equipment changes. It called this "state-of-the-market" technology and found it had the best payoff for everyone: employees, customers, and the company.

Layout. To avoid significant, ongoing disagreements, a holistic, systems view must be taken by all parties: IFM, MFG, and MPD. Most inconsistencies in priorities arise from not taking such an integrative approach.

The different manufacturing concepts presented throughout this book have great impact on all aspects of IFM: structure, layout, and materials handling. The choice of concept must be completely understood and its effects anticipated throughout plant design and new construction or modification. A continuing (internal) interaction must exist among the strategic decisions made about the facility's scale and scope, and the developmental site planning and plant structure and layout. The modular aspects of the facility itself can also be applied to the equipment, routings, and handling systems. The initial choice of equipment affects the ability to sustain the manufacturing concept a great deal more than does the capital budget, although money is not unimportant. Logical development of the site and related sites can also have an enormous impact on the internal layout of the various facilities.

Systemic analysis (See Chapter 5) should include the way product will flow, the way materials or parts will reach equipment or assembly lines for further processing, how waste and by-products are handled, and how people and equipment will interact for operations, changeovers, and maintenance. Storage points and volumes for materials, parts, consumable stores, tooling, and finished product must be factored into space requirements. Work and meeting areas for operating and support personnel should be tied into the way the plant will be operated.

What Else Do We Need to Know?

What follows is a summary of the key concepts and the jargon involved with the origins and the resolutions of conflicts within the three manufacturing processes modules:

Language of site: real estate, economics, financing, taxes, labor, leasing, contracts, etc.

Language of utilities: energy, water, sewers and drains, lighting, ventilation, trunks, spines, etc.

Language of equipment: capability, capacity, reliability, maintainability? Jigs, fixtures, tooling, gauges? FMS, autonomation?

Language of equipment operations: mean time between failures, changeovers, runtimes, bottlenecks, flow time.

Languages of regulation (OEO, OSHA, EPA): What constitutes regulatory compliance?

Languages of safety and health: for employees, community, and consumer.

Role of ergonomics in efficiency and safety (e.g., carpal tunnel syndrome).

Language of plant layout and materials handling flows in manufacturing: such new terms as automated storage and retrieval system (AS/RS), automated guided vehicle system (AGVS), "Pull" or kanban systems.

Language of controlled environments: positive and negative pressures, air curtains, filters, remote control of hazardous materials, clean room.

Language of maintenance planning and control (responsibility shared with production?): consumable tooling, supplies (MRO), reactive, preventive, predictive, total productive maintenance, employee-centered maintenance.

Performance Measures

Significant differences in the criteria by which "the boss" evaluates performance can generate development conflicts if based on the traditional silo orientation of individual SD team members. Senior management eventually must establish shared performance measures that can guide the whole company's efforts and do away with existing traditions.

Examples

A new maintenance criterion is *equipment availability*. We all understand that the best time to take the car in for overhaul is when we will not need it for a while. The plant equivalent is doing preventive and predictive maintenance in coordination with scheduling so that maintenance work is planned when the equipment will not be in demand. The result is equipment that is available and operative when called for.

Scheduling 90 percent of all maintenance—versus reacting to breakdowns—means that several subsidiary performance measures are quite stable: mean time to repair (MTTR) will be quite predictable when most repair work can be planned. Mean time between failures (MTBF) is improved by two other activities: efforts to improve product quality and reliability and the effects of predictive maintenance. Mean response time (MRT) will improve as fewer breakdowns occur and operators do more of the routine equipment monitoring and maintenance. More warmth in the relationships between maintenance and production will be the result.

Budgets. Some traditional budgeting methods lead to precisely the wrong actions by functional management. Many readers are familiar with the negative impacts of the manufacturing budget (absorb the overhead; make lots of inventory) and the end-of-the-month rush. Significant concept changes in the plant, new-product assignments, or the need for more training, can cause overruns unless the changes are part of the original plan. Site management must work these through and develop the equivalent of flexible budgets for all departments.

In some cases, present budgets do not even reflect the customer expectations that will result in more business—or in order cancellations! Quality, on-time delivery, field service response time, flexibility to unexpected customer needs, and supply service levels are examples of non-budget items that can make or break the site or the company. All people will respond to how performance is actually measured, not how it is described in press releases.

Safety Records. The statistics of safety represent a frequently reported performance measure: number of lost-time accidents, lost-time hours, and the lost-time ratio (accident lost hours divided by total hours worked in a time period). Such data also appear in regulatory inspection reports. IFM is the most frequent custodian of both records and related investigations.

In practice, an effective safety program increases equipment availability, up-time, and production, and lowers compensation claims and insurance premiums. Of even more significance, however, is the effect of plant safety (or its lack) on the morale, productivity, and interest people in the plant show toward improvement, including working safely. Further, lost-time accidents can deprive a team and the plant of skills and ideas that make a difference to overall plant performance.

The other side of this coin, however, is that of influencing people to stay on the job when their performance might suffer and they might be at greater risk of injury, to try to keep "the statistics looking good." This may put human resources, IFM (through insurance and safety), and manufacturing at odds. Resolution is important for both short- and long-term plant success.

Team Interrelationships

The site development team is empowered to get the job done. The three traditional measures—"on time, in budget, meets specs"—are augmented by many qualitative project performance attributes frequently more important than even the *sum* of all the traditional measures. Mission success and customer satisfaction are but two such measures.

The following will help in recognizing the potential for conflict in project work.

CPM and PERT. These are tools to help both planning and control. Each can be used to predict (simulate) the effect of changes of task completion on the project as a whole. Each can examine the effect of "crashing" critical-path tasks in time and cost terms. Each can assess the potential for lot splitting and overlapping to gain some time, as well as examine the risks of doing so.

But no tool can think for the team. And when dealing with qualitative aspects of the project, PERT/CPM can only provide partial responses to the truly important questions. Pressures to delay or to hurry up or even to stay on track are almost certain to be present throughout the project life.

Timing and Production Needs. The marketing (product-process development) goal is to beat the competition into production. Major customers may have been advised of product availability. Promotional campaigns may have been launched. What are the impacts of these pressures, and what amount of crashing is justified? How important is sustainable quality to the long-term customer relationships?

The converse may also be true: The capacity is not needed on the original schedule or there is uncertainty about the specified technology being able to provide a sustainable advantage. What do these do to the project plan? Staffing levels? Authorized contracts and shipment dates?

Examples

Forecasts. The proposed site of another of TVA's power-generating nuclear reactors—a $4 billion excavation and foundation structure at Hartsville, Tennessee—stands as a monument to forecasting error. The project had to be shut down, not because of environmental activism or regulatory intervention, but because the forecasts of power demand were made before the "oil shock" triggered massive energy conservation efforts nationwide. Then TVA chairman Marvin Runyon "bit the bullet" and moved TVA back toward its original mission, with short-term repercussions like the Hartsville shutdown as the necessary consequence. Today, more than 10 years later, as aggregate demand forecasts again appear to justify more power stations, TVA may reactivate the site.

Team Composition and Key Player Availability. The various aerospace and military programs developed from the 1960s to the 1980s encountered the problems of scarce skills and not-yet-invented equipment and thus suffered from bidding wars and crashing. The availability of key players and of experienced multifunctional team breadth can greatly affect all aspects of project success. Within the company, these scarce resources are sometimes the subject of executive committee debate. Even after allocations have been made, corporate priorities and customer needs may change. With scarcity a factor, project success may depend as much on negotiating skills as on project management!

Management. Personnel scarcity, company philosophy, other projects, and the length of the planning horizon for the project under consideration may all affect the decision about full-time versus collateral-assignment staff. Generally, large projects need a cadre (core) staff, augmented as needed by other specialists. But is *large* the same as *important* in terms of personnel assignments?

Logistics

Integration with Suppliers, Production, Distribution. Industrial site development is really a systems project. The entire logistics chain must be recognized for its importance to the facility being modified or constructed. Then the materials team of suppliers, converters and assemblers, and distributors, along with the transportation suppliers and customers, will all de facto be team members, whether assigned or not. Some simultaneous engineering teams also include the paying customer to help achieve practical optimization for the system. Obviously, many independent viewpoints will be expressed and will need resolution in favor of the system.

Customers and Products

Integration with Field Service and Customer Needs. In some respects, the systems project covers this potential conflict area. But understanding the form, packaging, volume, and reliability of manufactured products from the customer standpoint can profoundly affect how the facility develops.

CIRM Support Elements and World-Class Manufacturing Issues

Finance. This becomes the battleground for budgets and performance assessment. To be a world-class manufacturer (WCM), all facilities must be able to at least *meet* the globally determined competitive benchmarks for comparable products or services. On balance, site development teams must include the executive visioning inputs that set the WCM targets. These, rather than budget, must guide what the plant becomes.

Financial versus Regulatory Minimums and Maximums. Yet another conflict area is that of minimum compliance versus preparation of the facility to remain in compliance for a number of years. Here the LLCC issue is probably dominant. Relative discounted costs are only one comparator, because the costs of grayfield project management, production interruption, and retrofitting are often not included.

The problem is that because both regulations and technology are dynamic, today's solutions may be rendered ineffective tomorrow by changes in either one. The environmental engineering approach may be appropriate here, but can we afford the time and effort *now*?

Quality Issues. The conjunction of interdependent issues addressed in the CIRM approach is most evident in the implementation of what is loosely called *total quality.* Consider the marketing and product development aspects: perceived value on the one hand and price on the other; distinctive product attributes and production costs; innovation in design and ease of manufacture; and new, as contrasted with proven, production technology.

Human Resources. Quality has design and process dimensions; it also has human ones. How shall the integration among product, process, and people be achieved? What are the direct and indirect work force consequences of a design decision and a cost target? If we are substituting capital equipment for direct hands-on production activities, what is the new set of competencies that must be possessed by operating employees? By programmers, support staff, and maintenance technicians? How shall these be recruited and developed to become a consciously involved part of the entire manufacturing system?

Manufacturing Processes

Equipment Selection. What capacity, capability, and versatility are required for the products or families of products contemplated? In what ways are these reflected in the processes and equipment selected? How does equipment capacity affect both capital and operating cost?

What are the cost and performance trade-offs for the current (versus future) plant charter? What trade-offs exist in the product cost elements?

How are the labor, materials, and equipment elements of manufactured cost affected by such choices? How can these choices be related to the manufacturing concept, with its strong implications for work force skills and task assignment?

What are the effects of process and equipment choices on tooling and fixturing? If multiple machines are chosen over fewer "mega-machines," as driven by the manufacturing concept, what impact does this have on structure and layout? On materials handling and utilities? Personnel? Maintenance? Plant changes over time?

These complex relationships are shown in Figure 2–4 and Tables 2–2 and 2–3.

Process or Product Equipment? The nitty-gritty of planning departments and the various kinds of layouts (and materials flow) are natural combinations for each classification. The manufacturing concept (how the plant will be run) and the kind of plant focus (products or

FIGURE 2–4
Manufacturing Line Operations

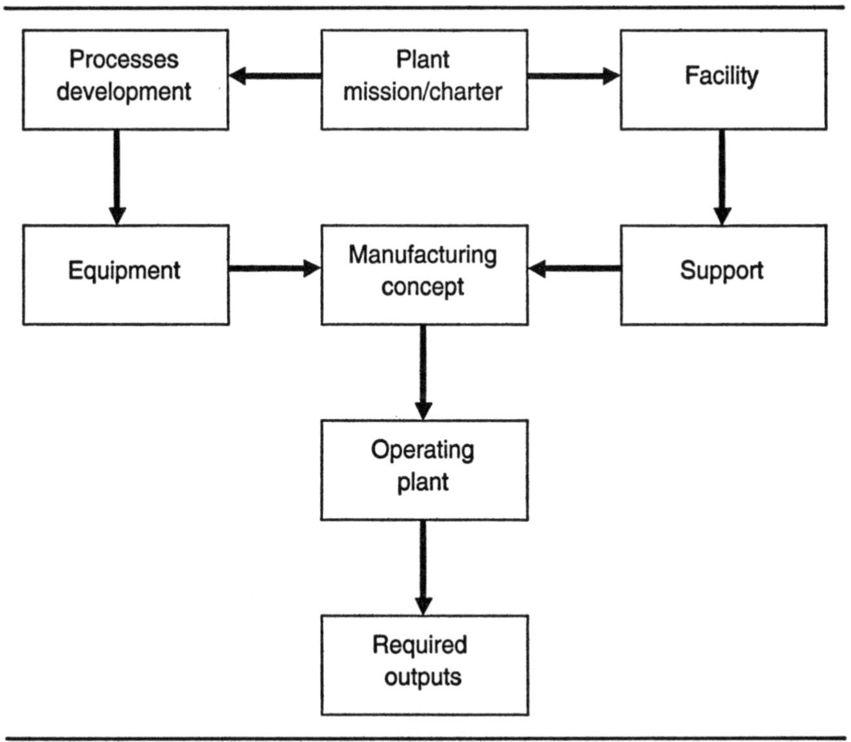

Source: Burnham and Natarajan, *Manufacturing Processes*, Student Guide (Falls Church, Va.: APICS). Reprinted with permission of the publisher.

processes) must be carefully thought through by the team. There are clearly many combinations of choices for equipment.

The modern plant is often a hybrid, with large *process* investments (brewing, pharmaceuticals, chemicals, petroleum, steel, aluminum), where these are appropriate, and then dedicated *product* equipment to carry the conversion forward to meet a specific customer need. This also applies where the minimum purchasable units of equipment capacity may make dedication to a single product uneconomical.

Note the opportunities for conflict. Cost economics often require process centers, while customer focus will suggest a product line orientation. Good systems analysis and materials flow studies comparing the probable production mix and equipment economics can help achieve a rational result.

TABLE 2–2
Manufacturing Line Operations: Considerations

Equipment	*Manufacturing Concept*	*People*
Capacity Per unit Total	Dedicated Centralized	Number of operators Skills
Capability Reliability Repeatability	Computer Managed (CIM) GT, manufacturing cells	Operator-centered Quality Maintenance
Versatility Inherent Adaptive	Computer-managed (FMS) Dedicated Quick-change tooling	Operator-centered Changeover Multiskilled shift (cross-training)

Source: Burnham and Natarajan, *Manufacturing Processes,* Student Guide (Falls Church, Va.: APICS). Reprinted with permission of the publisher.

Manufacturing Concept and Expansion Phases. Addressing these issues as part of contingency planning certainly reduces the conflict potential. But the question of what (capacity, equipment, or layout) should be part of the first phase is certain to arise unless the charter/mission statement is ironclad. And if it is, *should* it be?

Capacity and Capability. These equipment characteristics are performance-related. Should they become *concept*-related, the new system may not call for the same machinery. And where in the plant the equipment is used may affect the required capability. (Note the connections with "quality" above.)

Here, the LLCC criterion, modularity, and the need for flexibility may work together. In a product orientation, small machines must match with line requirements. In a process orientation, large machines must meet the most stringent requirements of both capacity and capability for that portion of the plant production assigned to them.

"Side" issues in equipment choice include features that answer how the equipment will fit into the overall operation. The fail-safe features of equipment that will stop the production run if any defects are detected may be either essential or unimportant, depending on the required volume and rated capacity. Single-minute-exchange-of-dies (SMED) is needed if the equipment is to serve many products, but is unimportant for only one.

TABLE 2–3
Manufacturing Line Operations: Relationships

People	*Layout*	*Materials Handling*
Specialized training	Process (department)	Spine and flexible material handling vehicles
Multiskilled teams	Product (line, group technology)	Lines (multiple spines or modules)
Multifunction teams	Routings CAPP	Cells Point-of-use delivery Point-of-use manufacturing

Source: Burnham and Natarajan, *Manufacturing Processes,* Student Guide (Falls Church, Va.: APICS). Reprinted with permission of the publisher.

Reliability. To judge the reliability of a proposed purchase, the maintenance database should include performance histories on the plant's machinery and the databases from other plants. The experience of the operators and the maintenance staff will be important if they have had opportunities to work with the proposed equipment. The recommended maintenance intervals and the spares kit from the supplier can give some indications. Again, note the quality connection.

The proposed use should indicate what reliability is desirable. And if experience with the equipment is lacking, reliability assurances may be sought contractually from the supplier. For major equipment, IFM may review the supplier's reliability studies or develop them independently. So where are the conflicts here?

IFM may look at LLCC from both acquisition and repair cost standpoints. MFG wants maximum uptime; MPD may want key performance features, though the machinery may need specialized operator and maintenance attention. Both the required skills and the number of maintenance technicians will be affected by the choices made, so HRD is also concerned.

If multiple smaller machinery units have been chosen and organized into cells, can the availability be assessed by looking at the collection of all machines? This might mean that a single excess unit could mean 100 percent availability for all.

Maintainability. Ease and frequency of maintenance can affect LLCC. Easily obtained parts, simplicity of design and assembly, and comonality of parts across different models from the same manufacturer are other features to be considered. The IFM members of the team will probably have the most information on this aspect, although process design and development is done through manufacturing engineering (MPD). The goal is to be sure that maintainability is part of the equipment choice process.

Environment. The operating environment can significantly affect equipment choice, or vice versa. From paint booths to semiconductor clean rooms, there must be consistency between process, equipment, and needed results.

MANAGING NEW FACILITIES ACQUISITION

A multitude of considerations affect the planning and commissioning of a new facility. When all the smoke has cleared away, what remains must reflect the best efforts of the site development team to represent the logistics, marketing, and manufacturing strategy of the corporation.

New or Revised Mission or Charter?

Why is a new facility needed?: demand expansion? new-product introduction? natural resources depletion? facility obsolescence (it is cheaper to start over than to retrofit)? new market or new technology?

The resulting charter and mission statement must provide the guidance the SD team will use to develop the plan. If consensus is difficult to achieve, the charter must decide for the company's betterment among all the differing view points, time horizons, and criteria that will be applied by the group of interested parties.

New Manufacturing Concepts

Manufacturing management procedures are unlikely to be detailed in the charter. Because it is vital to include how the plant is to be managed from the outset of the plan, *concept* requires another series of debates. Some of the possibilities with the greatest impact are discussed below.

Group Technology

Analysis of materials, processes, and applications of the variety of parts and products assigned can yield commonalities which—perhaps augmented by weight-size, production volume, and special features—can suggest that machines be grouped together to focus on production matching the common aspects. This is a special type of focus—that of process—and it leads to smaller work cells, with many of the independent characteristics of a plant-within-a-plant (PWP): a dedicated, flexible, multiskilled operator team; dedicated tooling, fixturing, equipment, and materials handling support; and little paperwork. There are clear impacts on cost accounting, budgeting, materials management, layout, staffing, quality control, and scheduling for the plant. The logic of group technology (GT) can also mandate the development of a manufacturing database. This, in turn, can motivate much greater standardization in part and product design.

Poka Yoke (Mistake Proofing)

In concept, poka yoke seeks to stop processes as soon as an out-of-tolerance condition exists. Carried to its logical conclusion, the design of both product and process will reflect mistake proofing so that defects will simply not occur. The formal execution of mistake proofing leads to autonomation. Where this does not take place in design, the continuous-improvement process will address ways to make poka yoke operational in the plant. Consider some of the interfunctional relations of technocrats and financially oriented participants: budget, product and process development, design target completion time, operating cost, training, scheduling, computer support, and manufacturing management.

Other New Concepts

Focus factories, cell manufacturing, self-managed work teams, synchronous-flow manufacturing, and the many variants of just-in-time can each present conflict situations for both planning and project management. At best, one can hope for creative chaos, as the actual construction period (perhaps 12 to 18 months) offers many opportunities for motivated functional specialists, prospective operating personnel, and team members to discover even better ways to organize the facility.

New Logistics Concepts

The global market and the recognition of logistics as a ''big system'' that includes manufacturing offer multiple opportunities for conflict. Marketing, purchasing, data processing, accounting, and manufacturing all have responsibility for aspects of the materials acquisition, transportation, conversion, distribution, and product sale leading to customer satisfaction.

Depending on company organization, either an experienced logistics person or a temporary representative from each functionally responsible area must be involved as the patterns of materials movement and the corresponding information system tracking are developed. These affect structure, layout, building design and orientation on the site, traffic patterns, and the macro-movements among sites and to the customer. Logistics is necessarily a companywide concern, and the development cycle is probably as long as that of the facilities which are thus tied together into a system.

Point of Use Delivery

From rack jobbers to supplier trucks unloading next to an assembly line, many opportunities exist to reduce handling, counting, and storing and to enhance the availability of parts and assemblies where and when needed. Access and layout must fit this need. Accounting, probably through information systems barcode and data capture, must find ways to stay informed of receipts and consumption.

Consolidation and Peddling

Common and contract truck carriers have for some years offered the road-based equivalent of the airline hub-and-spokes system. Carriers have developed economical means of gathering small loads, moving the collected mass over trunk line hauls to near the destinations, then essentially breaking bulk and moving the small loads to their site of use. This has become highly refined and optimized, under the concept of JIT, for large fleets of owned and contracted vehicles with supply and consumption points. Knowledge of the overall rhythm of the system helps with receiving and shipping-dock design, road patterns for truck movement, and—for bulk products like wood chips or paper fiber rolls (e.g., raw materials for Pampers)—the synchronization of rail traffic for inbound and outbound railcar strings. The facility and the internal materials handling system *must* fit with the logistic system or be adjusted to do so.

Total Productive Maintenance

The benefits of total productive maintenance (TPM) include better overall organizational functioning, less conflict, and more satisfaction for all the employees involved. The challenge is to make or obtain the commitment to undertake the several-year journey. TPM, like TQM, does not offer the immediate visible benefits of JIT implementation. Maintenance tradition is quite strong in many companies, and most software addresses planning and control rather than predictive/preventive maintenance improvement. Most greenfield projects will probably not begin with TPM but, over time, will move toward operator-centered maintenance and then TPM.

Reduced Design-to-Production Lead Times

The site development team, if established and supported, can bring a new or modified facility online early and with little turmoil and with competitive advantages for the firm. Shorter product design time is a great competitive advantage, and it is actively sought among progressive companies.

Lower Product and Facility Life Cycle Costs

Structuring the facility development phases to parallel the expected growth in demand and using both modularity and flexibility in the plant design will assure LLCC. Using the activity based costing approach will permit effective evaluation of costs and benefits of proposed new products, while not miscosting those already assigned. This will tend to discourage misassignments and help maintain the plant's product or family focus and related manufacturing concepts.

If an integrated approach is used, and thorough contingency planning takes place, the company will achieve the lowest-cost *capacity* strategy as well as LLCC for the plants.

Robustness, Customer Expectations, and Quality

Robustness means tolerance for variation while meeting quality mandates. The quality dimension is woven into the tapestry of the integrative facility development. Equipment procurement, layout, installation, maintenance, and modification to meet the needs of manufacturing are all included. The payoff is a satisfied paying customer, and the method is the proper combination of all variables to meet it.

Decentralization

The difficulties of effectively managing large, complex, centralized organizations has led to focus (reduced diversity of product and task) and regionalization (campus sites, groups of module manufacturers serving an assembly complex). This decentralization is consistent with the trends toward part and process "make" rather than "buy" and toward employee empowerment.

All of these trends represent an effort to put performance responsibility and accountability at the lowest possible level in the global organization. The existence of excellent worldwide communications and the global supplier-customer emphasis make such decentralization both satisfying and necessary.

LESSONS FOR THE INTEGRATED RESOURCES MANAGER

Shared understanding of the overall strategy, of which various plant-level plans and activities are a part, is a principal benefit of the site development concept. This understanding and recognition of the need to work closely together as a team (*and* with the work force and other staff) will have positive effects throughout the plant organization. It will show up as:

- Shorter implementation cycles.
- Lower costs.
- Higher quality.
- Organizational optimization.
- Reduced conflict and greater job satisfaction.

There will be choices of equipment with characteristics corresponding with the needs of:

- Reliability (MTBF).
- Maintainability (MTTR).
- Capacity.
- Capability.

Finally, the acquisition of new facilities (greenfield) or modification of existing ones (grayfield) will be meshed with time-phased strategic site development and will anticipate future missions and configuration needs.

REFERENCES

APICS: The Educational Society for Resources Managment. "1993 CIRM Study Guide." Falls Church, Va.: APICS, 1992.

Burnham, John M., and Ramachandran (Nat) Natarajan. *Manufacturing Processes,* Student Guide. Falls Church, Va.: APICS, 1992.

———. *Manufacturing Processes, Instructor Guide*. Falls Church, Va.: APICS, 1992.

Bell, Robert R., and John M. Burnham. *Managing Productivity and Change*. Cincinnati: South-Western Publishing Co., 1991.

CHAPTER THREE

FACILITIES MANAGEMENT AND CIRM

The earliest major conflict likely to arise within the group is in the context of plant expansion (grayfield modification) or new site development (greenfield), the latter often involving new products or customers. At issue are the many factors of investment, time required for completion, interference with current production, time for and cost of site-location studies, resource availability, ease and cost of transportation, and regulatory constraints.

BALANCING THE ELEMENTS

The texts in this CIRM series have a common theme: the relatedness not only of the 13 topics they cover but of those topics of interest to the many internal and external customers and suppliers to the resources system. Beyond these are various "influential others," such as shareholders, industry competitors, substitute industry competitors, and regulators. Figure 1–1 suggested the interdependencies that exist in fulfillment of IFM's many roles.

To even begin to consider the means through which the many CIRM elements can come together in carrying out IFM strategic (representing, or "catalyst"), tactical (enabling, or "glue"), and operational (support, or "lubricant") assignments, it is necessary to understand the concerns of each of these partners in the management of the enterprise. And each partner must be concerned because any failure affects them all. The general solution to this challenge is the use of teams, and the integrative resources manager will be an effective team member.

To glimpse how these functional strategic elements can be balanced to help achieve the goals of the enterprise, consider Lincoln Electric, an outstanding company with a clear strategy of winning through focus on the customer and effective use of manufacturing as a "weapon."

LINCOLN ELECTRIC

Lincoln Electric, an arc welding equipment maker, is a premier company that dominates its market niche. Lincoln's strategy? Excellence in manufacturing. Everyone in the work force actively seeks ways to improve the product and the way it is made. Lincoln's rate of welding machines produced per employee is more than twice that of its nearest competitors.

How is this achieved? The company's manufacturing strategy addresses a combination of highly complementary elements that define the requirements for maintaining competitive advantage. Plant size and location are optimal for *all* production and distribution functions (*representing*). Equipment is upgraded, modified in-house, and layouts are reorganized to match manufacturing needs. Quality is absolute (*enabling*). The blend of work force involvement and motivation, support for improvement, piece rate incentives, and very substantial annual bonuses to the entire workforce lead to Lincoln's outstanding results (*supporting*). The aim of facilities operations is "Never having to say you're sorry," with excellent maintenance, flexibility, and a safe, healthful work environment.

Note the balance among the elements supporting the Lincoln strategy. People, equipment, quality, and continuous improvement—these, in combination, quite literally obliterate the competition. The specification of manufacturing systems, controls, and operations can fulfill the policy guidelines and attain productivity, quality, service, and return-on-investment goals for the firm. Facilities are key to carrying out the strategy. And while primary interactions are among facilities, manufacturing, and process design personnel, it is clear that many other elements are involved: marketing, product development, supplier, equipment manufacturer, and human resources personnel being the most obvious.

Figure 3–1 suggests some specific relationships among the 12 other CIRM subjects and integrative facilities management.

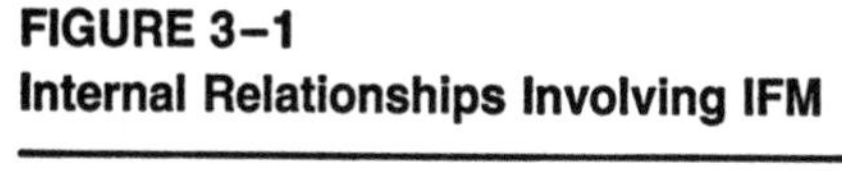

FIGURE 3–1
Internal Relationships Involving IFM

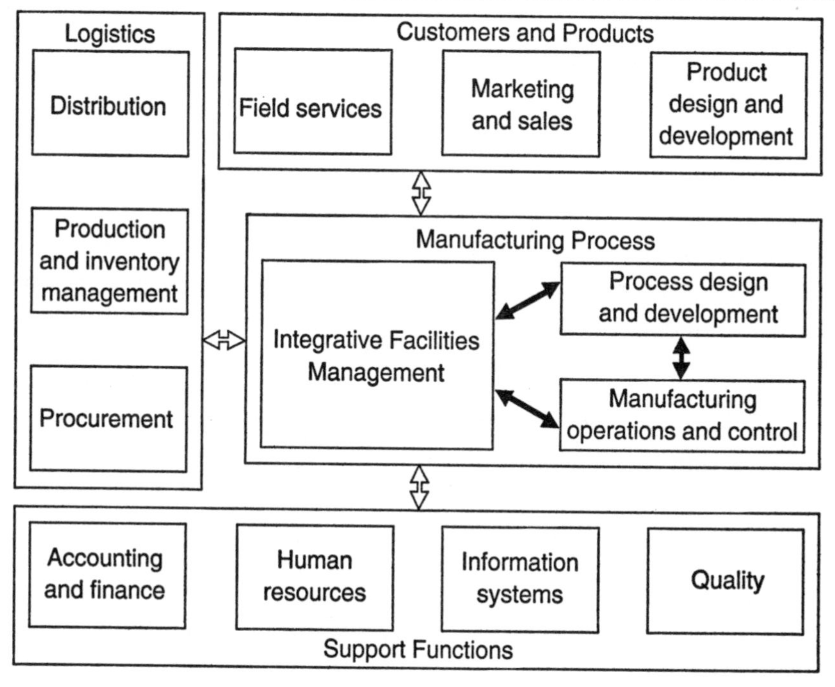

CIRM AND IFM

The Lincoln Electric example can be used to show relationships across the board.

Customers and Products

Lincoln's *field service* is best exemplified by "the Maytag repairman" snoozing away a lonely existence waiting for the repair call that never comes. But field service is also a strong marketing tool, helping the customer to help Lincoln become indispensible! The service facility must provide excellent support and be equipped with adequate stock levels of replacement parts, work benches, vehicle parking, and maintenance bays (*and* cots for the unneeded repair personnel!).

Marketing and sales provide product acceptance reports and sales forecasts, both of value to IFM forward planning activities (new or modified products, changes in capacity or layout). The typical Lincoln salesperson is a graduate engineer and has learned how to weld before going into the field to work with customers. Many customer technical problems are solved this way, linking the customer with Lincoln. Sales and marketing offices reflect the company's priorities in meeting customer expectations.

Product design and development frequently interacts with manufacturing process design and development, and IFM must be a partner in the generation of facilities responses. Field feedback from both service and sales helps with modifications of products and with development of new ones. And as new, or additional, products or services are conceived and moved forward toward availability, IFM must be proactively involved through the various site development programs at manufacturing, distribution, and services facilities. Charters may need revision, site plans changed, and location choices analyzed in light of the new proposals.

Manufacturing Processes

Changes in technology, in products, in design or processes, and in operations are frequent in customer-driven companies, and maintaining a competitive advantage makes them a top priority at Lincoln. These three manufacturing functions—process design, facilities design, and manufacturing operations—must work closely together in maintaining the entity that produces the welders and rods bearing the Lincoln name. In general, new products can require new processes (and conversely, new processes can "enable" new products), and either or both can present the need to analyze the total manufacturing system from the standpoint of the adequacy of the existing facilities to support the new mission.

Facilities

The original Cleveland plant grew, and Lincoln constructed another plant nearby and acquired other companies and their facilities in the same manner over the years. IFM supported all these acquired facilities in the same manner as the original Lincoln facilities. In all cases, the development of the production organization has been integral with its equipment, the plant structure and layout, and the concepts that will govern manufacturing operations. Lack of equipment standardization is a potential problem in acquired facilities. Lincoln found that most of the equipment was similar but not being used as effectively as in Cleveland. The manufacturing concept affects layout, machine groupings, required floor areas, the overall materials movement pattern, and therefore the materials-handling aisles and equipment. It can also affect structure and the development of ventilation systems for flexibility and safety.

Lincoln's offices are "functional," with no trimmings. Old desks still serve, and no "gimmicks" exist. Until recently, the sole private office was that of the chairman, and it was carpeted only to provide visitors some comfort and to serve as a conference room.

Manufacturing

Lincoln's incentive system rewards employee productivity. Observers at Lincoln note that the whole floor seems to have evolved into JIT: The layout supports almost continuous flow, and there is little inventory despite the variety of models and options needed to meet customer expectations. JIT, TQM, and other advanced manufacturing concepts require that IFM both respond through *support* for needed changes and *enable* these changes through detailed planning for both greenfield construction and grayfield modifications.

Process Development

Equipment changes are not made to be fancy, but aim at improvement. Old equipment operates alongside new machines, and both are maintained in quality-capable state. Appropriate technology is aimed at reducing time, rework, materials consumption, materials handling, and travel distances as the product fabrication and assembly take place on the way through the plants.

Support Functions

Lincoln quality is legendary. The work force carries out quality checking as a part of operations. Equipment is maintained in "capable" condition. Maintenance procedures are designed and carried out by operators and by IFM personnel.

IFM at Lincoln needs considerable *information support* for each site, both technical (layout, equipment specifications, maintenance history, utilities data) and administrative (safety, environmental regulatory compliance, maintenance inventory). Access to CAD/CAM technology for products can help customers and products, manufacturing processes in general, and especially IFM to keep plant drawings up-to-date and to enable effective planning.

The Lincoln organization is one of decentralized operations and strong incentive systems. *Human resources* must recruit and orient personnel to increase Lincoln's capacity, with the skills to support Lincoln's way of doing business and the desire to be self-managed. Because all employees become permanent after two years, the selection and development processes are very important. For example, maintenance and changeovers are carried out by operators as much as possible. This requires operator skills development, as well as formal and informal training by IFM specialists. Support for equipment modifications to improve operator performance is also sought. This trend in business promotes employee involvement and job enrichment.

Finance and accounting and IFM have ongoing dialogues by the nature of their tasks. IFM generates capital cost estimates and justifications for investments in facility and equipment assets. With the shorter product life cycles and the open, changing manufacturing environment, policy issues involving facilities and equipment costing must be resolved for the long-term benefit of the company and its products. Hurdle rates, the way to implement long-term capacity strategy, and the time phasing of capital costs to fulfill the site charter but stay within capital spending limits all require considerable discussion by the site development team.

Logistics

The three elements of the logistics system are procurement, production and inventory management, and distribution. For any JIT plant, and certainly for Lincoln, all three must be coordinated to achieve smooth, lowest-cost flow. And there are very significant interactions among the various elements of the logistics system and IFM.

Procurement

The quantity and frequency of materials or components delivery affect receiving docks and apron size. Point-of-use storage will affect floor space and the means of access by the supplier to the storage area. Point-of-use manufacturing will affect the nature, weight, and bulk of materials delivered. At Lincoln, Procurement is constantly seeking ways to manage the inventory investment while achieving high service levels.

Production and Inventory Management

Internal to the plant, the manufacturing infrastructure influences many IFM decisions. Manufacturing cells require different materials handling, storage, lighting arrangements, and aisleways than do assembly line configurations. Machines grouped by technology rather than by process also need unique IFM planning. Lincoln's small-lot manufacture (a little of everything, every day) calls for different (and generally more compact) layout than does large-lot manufacture, which needs considerable space for work-in-process and bulk materials handling. Kanban (pull) scheduling calls for much less data collection and display and much more visible control signals. Procedures and organizational decisions greatly affect IFM because the communications, *system* (though not *what* is communicated) is generally assigned there for installation and maintenance.

Distribution

The use of pull scheduling in manufacturing will affect shipping quantities, numbers and sizes of vehicles, and probably the facility size, receiving system, and materials-handling methods in the warehouses and distribution centers. Overall, facility locations, transport modes, and the means for meeting customer demands must lead to highly integrative choices at both representing and enabling levels.

The many different relations that affect IFM call for a site development team, with changing makeup and purposes depending on the planning horizon and the nature of the developments being contemplated. This will become evident as the system that leads to a greenfield plant (and, later, to the many grayfield modifications to keep it current) is described.

IFM AND CAPACITY DEVELOPMENT

Corporate management is responsible to the board of directors and through it to the shareholders. Thus, as a practical matter, the decisions that eventually lead to a new facility are ordered hierarchically, with the commitment of major corporate funds reserved to the executives accountable for their wise use. Depending on the nature of the business, the degree of capital intensiveness, the market targets, and the kind of facility being contemplated, the process may take months or decades, but typically from two to five years. The term cycle is used because of the recursive nature of the processes typically followed. This hierarchical set of planning activities is suggested in Figure 3–2.

As the strategic need for additional capacity is examined, IFM is active in the representing role as policies on market share are balanced

FIGURE 3–2
"Drivers" of Integrative Facilities Management

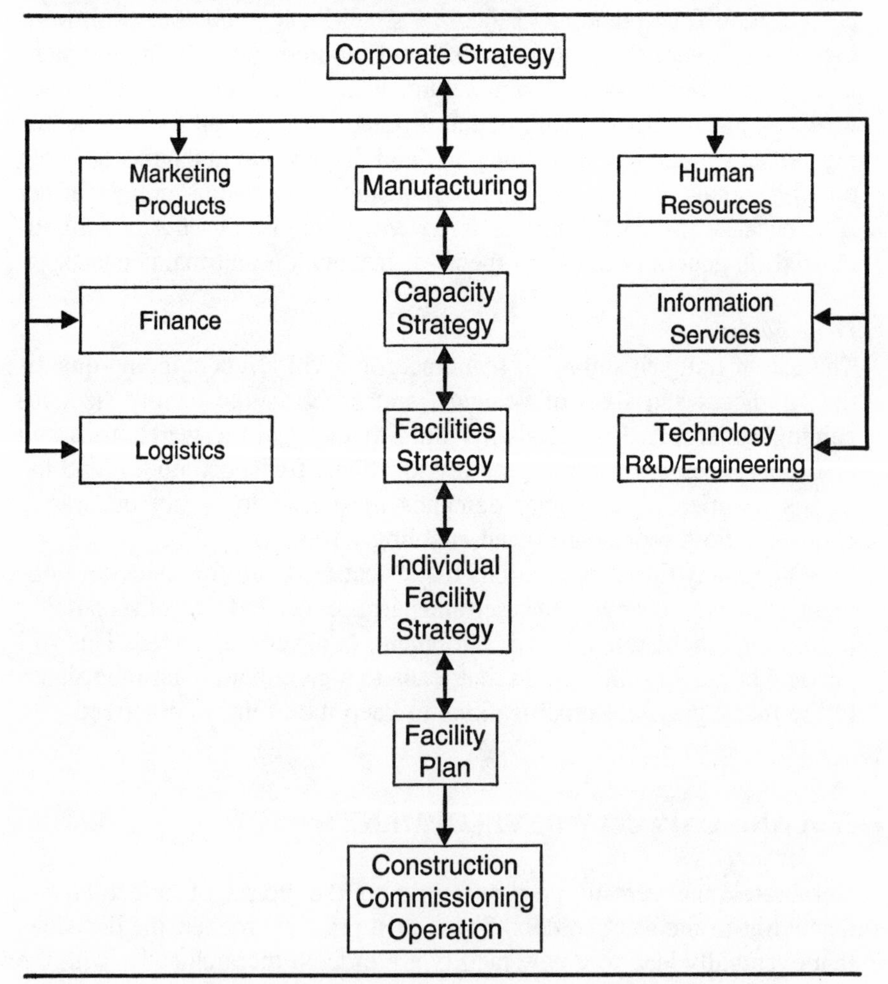

Source: John M. Burnham and R. "Nat" Natarajan, *Manufacturing Processes,* Student Guide (Falls Church, Va.: APICS, 1992). Reprinted with the permission of the publisher.

against the costs and time-based revenues new capacity can deliver. Various combinations of new and existing sites, market shifts, competitor locations and capacity, and demand predictability are used to evolve a facilities strategy. This, in turn, drives the development of individual site missions and charters, when IFM begins an enabling role. Each plant charter can be converted into a time-phased program of construction (or expansion or both) and preliminary and detailed facility plans developed. These plans, like the capacity and facility strategies, are examined recursively against the best information available at the longer-horizon level. Thus, while the traditional planning process originates "above," with directions passed downward for detailing or execution, this simply is not sufficient in the dynamic global world. If the appropriate multilevel, multifunctional teams are involved in this planning process, the teams effect the integration and recursive looping suggested by the double-ended arrows shown in Figure 3–1, designed to summarize this capacity and facilities process.

Capacity Strategy

A firm's capacity strategy is a deliberate pattern of decisions, over time, for achieving the capacity goals of the firm. These decisions are long-term and far-reaching in their effects and will lead to a facilities strategy affecting all of manufacturing and distribution.

A capacity strategy reflects the company's efforts to deliver quality products and services meeting customers' desired quantities and timing. Many factors bear on such policy development, including economics, technology, and the related capacity increments, competitor practices, customer expectations, and distribution/production location trade-offs.

The nature of the business and its products radically affects a feasible capacity strategy. A firm requiring long lead time, high capital equipment items, or specialized facilities must live with the results of capacity decisions that reach far beyond the forecast horizon for demand. Papermaking, oil refining, power-generating facilities (public utilities), and other process businesses require years to achieve significant capacity expansion. Facilities driven by such capital and lead time requirements suffer yet another penalty: They must be very intensively utilized (perhaps upward of 90 percent), to earn profit for the company.

These decisions become public knowledge once engineering or construction firms or equipment suppliers have been contacted or siting alternatives explored. The company seeking to gain market share, then, must act in a very public fashion, announcing plant expansions and commitment to the customer while perhaps expecting a corresponding action from one or another competing company. However, these characteristics provide some protection, constituting a barrier to new entry because of the financial risks attendant on entering such an industry without an established reputation and market. Signaling one's intention to expand thus preempts moves by companies "just thinking about" a similar project. Nucor moved very rapidly to build and activate the heavy shape steel plant in Blytheville to discourage competitors from emulation. Because of the perception of greater technology risk, Nucor's Crawfordsville thin sheet plant was not immediately copied, but other steel makers have been studying both the plant and its product.

Process businesses must examine the significant financial benefits of building a large facility. This "chunking" can save a great deal of money and provide barrier advantages. But deep pockets are also needed, because below-scale operating levels carry a financial penalty in underabsorption on the one hand and production inefficiencies on the other. As a matter of policy, a number of chemical companies choose to anticipate demand and be prepared for it, rather than attempt, with a long lead time and expensive facilities, to "chase" it. Just-in-time philosophy and practice has introduced another variable into the capacity equation: using excess capacity to be more responsive to the customer without carrying a large amount of inventory. These "diseconomies of scale" can be identified through study of Figure 3–3.

Process plants have singularly consistent economic models that show the "running-full" nature of most such businesses. Doubling the plant size will bring the total production cost (fixed plus variable) down to perhaps 60 percent (the six-tenths rule) of the former plant size cost. The same plants find that reducing changeovers will significantly affect average cost, as in job shops and other discrete manufacturing. For both discrete and process manufacturing, the analysis is straightforward. Calculations use the familiar PVC (profit-volume-cost) analysis to yield the cost per unit at various volumes.

$$\text{Unit Cost} = \frac{\text{Capital Costs} + \text{Variable Costs}}{\text{Production Volume}}$$

SIGNIFICANCE OF CAPACITY STRATEGY

Given the dynamics of the marketplace, major strategic questions concerning capacity additions—How much at a time? Located where relative to markets (or competition)? When do we have the new capacity commissioned and on-stream?—have serious consequences. The kinds of choices made during the development of a capacity strategy must antic-

FIGURE 3–3
Relationship of Plant Size to Operating Costs

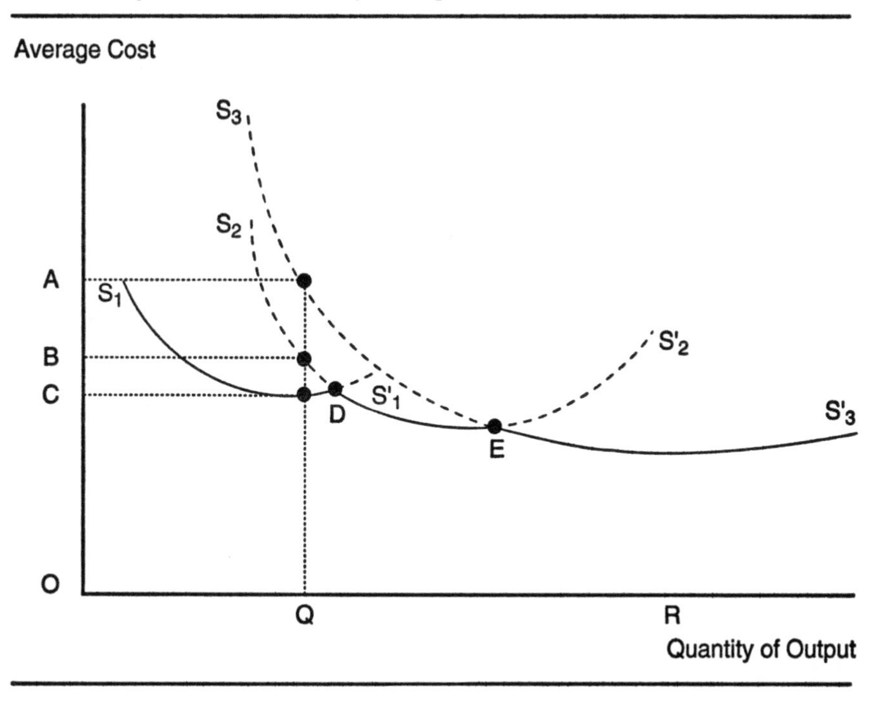

Source: John M. Burnham and R. "Nat" Natarajan, *Manufacturing Processes,* Student Guide (Falls Church, Va.: APICS, 1992). Reprinted with the permission of the publisher.

ipate, over a long planning horizon, trends in the global competitive arena and related markets.

The results of such time-phased capacity decisions can be evaluated in terms of the capacity cushion (reflecting perceived opportunity cost) and the form of that cushion (excess factors of production, extra equipment only, capital set/aside). The firm can evaluate capacity decisions for itself and for its competitors, helping to formalize policies related to customer service, lead times, demand forecasts, and the cost of excess physical capacity. These can also be used to make the capacity decisions on a rational, proactive basis by predicting competitor behavior and acting accordingly.

The capacity cushion can be calculated on the basis of probabilities: the trade-off between the cost of extra, unused capacity and the lost revenues from unsatisfied demand. This is the strategy of market share growth. It can be expressed in a simple formula yielding the desired probability:

$$\text{Critical Probability} = \frac{\text{Opportunity Cost} - \text{Excess Capacity Cost}}{\text{Opportunity Cost}}$$

Economics enter in when capital costs are significant, and the unit product cost is strongly affected by these fixed costs. The ability to increase production volume with little more than the direct (material + labor) production cost per unit leads to capacity policies affecting the *increment* of capacity addition. This chosen increment reflects the principal concern: cost minimization rather than revenue maximization.

Example

Caterpillar Company's capacity positioning has been consistent throughout the decades since World War II. Despite the big-ticket nature of the product, ''Cat'' seeks always to have extra capacity so that customers can obtain their first choice: new Cat equipment for their construction work. As global business expanded, assembly plants appeared overseas. Standard designs focused on quality and reliability, and automation helped achieve low manufacturing costs. Centralized capacity for machining provided tight control over quality and availability. The excess parts and assembly capacity provided both responsiveness and a large barrier to competitor entry. The revenues allowed continued investment in better technology and even more capacity. Harvard's Chris Bartlett calls this a perpetual motion machine, and it was effective for nearly four decades.

The difficulty inherent in this capacity strategy was that it assumed there would be no significant changes in either the intensity of demand or the nature of the customers and their financing and distribution preferences. Further, it assumed that there was no effective competition for Cat. As later events showed, this was simply not true, and so Komatsu has taken a significant market share away from Caterpillar.

Overall Facilities Strategy

The overall facilities strategy analyzes a single facility or group of related locations in the context of the total market for its or the group's products. Such a study leads to recommendations that specific capabilities be modified and estimates the timing, investment, and product/profit-related impacts of the undertaking.

The considerations that affect the overall facilities development plan include carrying out the capacity strategies (timing and amount of capacity relative to projected demand) and facilities strategies (sizes, locations, functional needs, product–market/process focus, and corporate philosophy) as part of policy deployment to enable definition of an individual facility or network. Once these overall considerations are evaluated, individual facility capabilities (and their missions and charters) can be defined and detailed.

The design of a manufacturing and distribution network is not a simple task. There are logistics concerns that seek lowest total delivered product cost across the network, where purchased parts, inbound transportation, conversion, and distribution costs are all considered as variable during the design stage. As shown, the size of the physical facility affects its operating cost. Transportation costs are affected by the available choices of modes (air, rail, truck, waterway) and competing carriers, the volume of goods to move through the system, and the availability of back-haul opportunities for the transportation company.

The environment and allowable operating levels can affect both scale and location decisions. Limited available land, restrictive zoning, degree of urbanization, and nearby natural land or wildlife preserves can each eliminate an economically and logistically ideal location from serious consideration. Location should always be a strategic management concern because of its close relationship with many other internal and external factors. Location can become an operational albatross if contingency planning is curtailed or ignored.

Example

Oster Corporation, pioneer manufacturer of kitchen blenders and other household products, chose to build five plants in Tennessee. Each plant was designed to fit the size of the community and local labor availability. Because Oster processes were "clean" and each would employ hundreds of Tennessee citizens, the company and the local industrial development boards were working toward the same objective.

Each of the plants had dedicated products and manufacturing processes. And each carried, by design, excess capacity for one or another of the key components or modules that went into the blenders—motors, castings, or plastic molding, for example—so that economies of scale were achieved and could be shared by the nearby "customer" plants. Thus, the network of plants, distributed across middle Tennessee, functioned as an integrated whole and minimized the many risks that a single large plant might have to face.

The decision to build a *maquiladora* just inside Mexico departed from this small network strategy. The goal—cost reduction—was put forward as justification for the action. Facilities specialists carried out the site development, supporting the planned labor-intensive assembly operations. One of the five original plants (in Dayton, Tennessee) was closed as the *maquiladora* came on stream.

Undesirable consequences resulted: transportation lines lengthened, special language skills were needed, and coordination problems multiplied. The network became distorted, and much of the cost saving evaporated through red tape and unreliable material movements. Inventories grew to compensate. The results are confidential, but Allegheny Corporation, the holding company for Sunbeam/Oster, is now in bankruptcy. A fall 1992 takeover by another company led to wholesale dismissals of many of the network management teams.

Significance of a Facilities Strategy

A facilities strategy can form a multiplant manufacturing system, or network, that can support the marketing, customer services, and profitability goals of the firm. This system takes on details embodying a manufacturing concept that makes sense for the economics, human skills, equipment, capital resources, and competitive positioning of the firm and its products. In addition, there are factors determined by the way the manufacturing facility will be operated (see Figure 3–4).

Just-in-time manufacturing will have a great effect on both inbound and outbound materials movement—not in terms of total annual volume, but in the size and frequency of shipments, how they are packaged, and where they are delivered. A focused factory will have less diversity, and probably a smaller facility altogether, than one built to support the entire product line. Manufacturing cells using group technology concepts require different physical arrangements and materials-handling support than do departmentalized job shops, even though the products created may be identical. The manufacturing concept will clearly affect staffing, skills required, work place design, management skills and requirements, and

FIGURE 3–4
Impacts of the Manufacturing Concept

Concept	*Diversity*	*Capital Needs*	*Management Tasks*	*Equipment*	*Work Force Skills*	*Supervisory Role*
Focused Factory	Dedicated to limited product line	Lower; better balance; higher utilization	Achieve balanced rate; match the customer needs; exception	Simple; matches narrow range of tasks; efficiency through balance	Repeating; learning; teamwork	Teams; self-regulating
Manufacturing Cells	Often dedicated to one item or a small family	Lower; better balance; higher utilization	Specify output; provide resources; cell small factory	May be special but matches state of small factory for item	Repeating; learning; multitasks; teamwork	Support team development; help with improvement
Group Technology Cells	Dedicated to parts family based on material processes and standards	Lower; common equipment; more flexibility	Support with tools; do analysis; influence design	Usually general; relatively simple; flexible	Multitask; multiskill; team learning	Support team development; help with improvements
Multi-department Factory (Traditional)	Full range of products and parts	Higher; hard to predict mix shifts; must handle full product range	Complex; dissipates efforts; reactive system	Some will be special, some general; often oversized; may run for only hours a week	Repeating process; Simple? Standards to be met?	Meet the standard; process direction

Source: Robert R. Bell and John M. Burnham, *Managing Productivity and Change* (Cincinnati: South-Western Publishing Co., 1991), Fig. 4–17, p. 97. Used with the permission of the publisher.

most operating practices and procedures. Lack of consideration for the needs and effects of such manufacturing infrastructure can lead to disastrous facilities mismatches. The site development team, including manufacturing representation, must wrestle with these alternatives until consensus is reached so that design features can be detailed with confidence.

The ultimate facility design is affected not only by the economics and technology of manufacturing but by the infrastructure of logistics, human resources and operating concepts, and procedures that are used to plan and track the work taking place. Capacity and facilities strategies are integrated with manufacturing processes design and production. Location, scale, customer proximity, cost, and regulation all affect land and facility acquisition, both for initial greenfield sites and for expansion. An optimal development will represent the best judgments of the strategists armed with data, information, and analytic intelligence. Without the careful creation of both manufacturing and capacity strategies and their projection over the long term, there is a tendency to be reactive and to lose control over the potential for a lasting competitive advantage.

An IFM Strategy System?

Heery International, a design services and construction management firm, has developed a formal process called *strategic facilities planning*. A significant aspect of this process is the formation of an executive steering group, a working group, and an information resources group made up of business unit managers. A real estate executive and various consultant resources complete the project organization. Each team is multifunctional, and the project organization is multilayered (see *Industrial Engineering*, June 1989, pp. 25–32).

A HEWLETT-PACKARD "CAMPUS"

Consider the example of Hewlett-Packard. HP is a leading electronic instrument and computer designer and producer. HP has pioneered the company facilities development as that of a network (or what HP calls a *campus*). A number of other computer firms—Tektronix, TI, AT&T, NCR, ROLM, and Tandem—have adapted these same basic concepts almost on a global basis, so what follows has come to represent an informal standard for electronics manufacturing.

The HP campus has evolved from a pattern of standardized facility designs for appearance and for research and development, prototyping, components, and assembly. The "neighborhood" needs good weather, population growth, a good professional environment, and a sizable population of potential employees. Land acquisition managers may buy a site for the entire campus at once, using the size of the projected *group* of facilities as guidance for the area required.

The usual HP facilities strategy initially locates a "parent" plant on the site and provides appropriate access, parking, and other support to handle a headquarters operation, some manufacturing, and considerable engineering activity. This parent plant will develop prototypes, debug them, and handle low-volume production of new products.

As projected demand materializes, one or more assembly plants will be located elsewhere on the campus site to take on high-volume manufacturing. A typical configuration will be a wagon wheel, with the parent at the hub and the dependent facilities on the spokes. As the demand increases further, perhaps a component plant can be justified to support the other facilities' needs. Figure 3–5 shows how this campus might develop.

FIGURE 3–5
An Electronics "Campus"

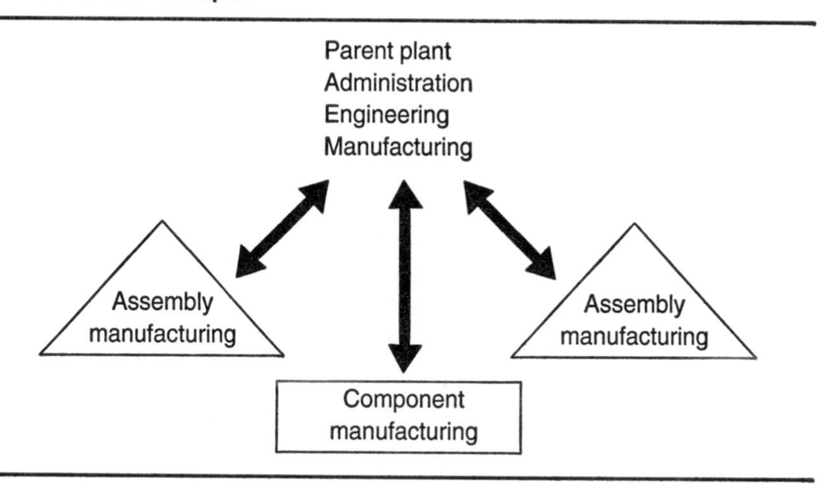

Source: William H. Hayes and Steven C. Wheelwright, *Restoring Our Competitive Edge: Competing Through Manufacturing* (New York: John Wiley & Sons, 1984), Fig. 4–9, p. 88. Used with the permission of the publisher.

Vive la Différence!

Keep in mind that for primary materials companies (ALCOA, Nucor, International Paper) the cost of transportation can be 15–30 percent of the cost of the product to the customer, so locations near raw materials, economical energy supplies, and/or major markets are extremely important. On the other hand, HP either makes or buys semiconductors, integrated chips, and microprocessor chips, none of which have either bulk or weight relative to their value. Even for the assembled product, there are transportation alternatives for HP that cannot be economically used by the steel products companies. Further, with the short electronic product life cycle, it is necessary for HP to develop a manufacturing complex that may have a 20-year expected life, with absolutely no idea of the products to be developed and made during the facility's lifetime!

When land acquisition, construction, and production are considered, other differences exist between Hewlett-Packard and a basic metals supplier like ALCOA. This offers another way of thinking about the importance of a capacity and facilities strategy and the many roles for IFM. Products that might be assigned to an HP campus over its lifetime can be contrasted with assignments at an ALCOA plant.

What differences will occur? While there will be exceptions, the ALCOA mills will principally serve *regional* markets because aluminum is heavy and transportation costs are significant when related to the value of the product. Aluminum product marketing must explicitly evaluate shipping economics. In most cases, HP will be a participant in *global* markets, with manufacturing, assembly, and distribution facilities.

HP must place great emphasis on being first-to-market with new products, and life cycles will be short. An HP campus will be closely tracking R&D activities within the company, as well as those of competitors. HP product marketing must identify customer needs that HP can serve and stimulate product configurations to meet them. For example, repackaging and miniaturization have developed an entirely new market for notebook and laptop computers that are powerful and lightweight and can travel easily. Product and production design may take place almost continuously. The facility choice(s) will depend on capacity and capability to build the new products.

ALCOA must provide high-quality products to established markets at competitive prices and will add products as demand develops and as the facility's capacity can accommodate them. R&D and product design/development for ALCOA will focus on metallurgy and casting capabilities and on modification of existing processes. These relationships are integral with the engineering, prototyping, and manufacturing tasks to serve the customer profitably.

Individual Facilities Strategy

The next stage in IFM develops and implements effective individual facilities strategies. These plans define the manufacturing and distribution network to achieve balance among current and future company locations in terms of resources and products and the supporting infrastructure for the strategic planning period (see Figure 3–6). In the Oster multiplant complex described earlier, it is clear that each of the five plants had specific needs to fulfill commitments to the other plants in the Tennessee complex and to meet corporate mission mandates related to other facilities and distribution locations.

With much the same emphases as for the corporation, a *charter* describes the plant's manufacturing strategy. Typical charter provisions include products and volumes, a mission statement defining key results areas and a forward life cycle scenario, process capabilities and capacities, development directions (all factors of production), and the performance targets associated with the key results areas.

The manufacturing concept shows how all factors of production are affected by the overall plant concept. There is a strong impact on how the plant site is developed for materials receipt and shipment; rate of flow through the plant; equipment choices, process design, and workforce skills; and organizational needs.

The Charter: What the Facility Is Empowered to Do

The proactive view of manufacturing as a competitive weapon requires that a charter should be set out at the beginning of facilities development and maintained and modified throughout the life of the facility. This

FIGURE 3–6
Facilities Strategy Development

Products/Life Cycle	Rate of Investment	Location Criteria
Growth Rates	System Network	Manufacturing Concept

Source: John M. Burnham and R. "Nat" Natarajan, *Manufacturing Processes,* Student Guide (Falls Church, Va.: APICS, 1992). Reprinted with the permission of the publisher.

facilities life cycle differs from—but is closely tied into—the industry and product life cycles of which it is a part. Start-up, expansion and modification, stability and refinement, and ultimately, downsizing and termination or redirection and rebirth, as life cycle phases, also require corresponding IFM activities. Elements of a charter are shown in Figure 3–7.

INDUSTRIAL FACILITY DEVELOPMENT

Practically speaking, IFM supports new construction or the expansion and modification of existing facilities. Annually in the United States, some $250 billion has been spent for these activities (an average of 8

FIGURE 3–7
A Plant Charter

Products and volumes (by item or product group)
- Volume capabilities
- Rank order of priorities required

What must the plant do well? (mission statement)
- Key leverage points and critical competitive factors
- General scenario for expansion/development (facility life cycle)

Process capabilities and capacities
- Capabilities by type of production process
- Capacities
- Changes over time

Development directions
- Plant and equipment
- Production planning and control
- Labor and staffing
- Engineering
- Organization and management

Specific objectives (by product and year)
- Unit costs
- Service level

Capacity utilization
- Scrap loss

Inventory investment
- Cost structure

Source: William H. Hayes and Steven C. Wheelwright, *Restoring Our Competitive Edge: Competing Through Manufacturing* (New York: John Wiley & Sons, 1984). Table 4–4, p. 100. Used with the permission of the publisher.

percent of GNP from 1955 through 1984), not including management time, consulting services, and so forth. Recent US government reports cite annual expenditures of over $1.2 trillion for nonresidential construction in 1991. Facilities construction is a highly significant portion of total corporate expense, and facilities assets amount to 38 percent of total capital for manufacturing firms, so careful and comprehensive facilities planning is a necessity.

Planning for the Plan

With increasingly proactive and strategic IFM emphasis, an organized approach to the planning process is needed. A typical engineering analysis sequence would begin with an audit, develop a plan, and then audit the new or modified facility in terms of ability to fulfill the requirements. A formal process with explicit enumeration of the what, why and how considerations is critical to strategic facilities development. The plan can then be developed and implemented.

"Triggers" for a facility plan can include either expansion or the need to reduce capacity. Market mix shifts can adversely affect a single site (particularly where transportation cost is significant), and major technology developments can force new study of both capacity and facility strategies. "Green" engineering that addresses by-products and waste can be the response to changes in either rule or enforcement timing of environmental or safety regulations. Finally, a facility plan can be developed to enhance customer focus and improve product competitiveness.

Facilities Planning

IFM's facilities-planning focus addresses both a greenfield new site and grayfield modification of an existing one. IFM must consider the plant construction, materials flow and plant layout, equipment selection and installation, and related design elements to support the capacity, process/product, and production requirements. Planning will include considerations of product mix and volume, and the quality requirements to result in a competitive product; the matching equipment characteristics, target manufacturing cost to be competitive, and the related technology(ies), workforce skills, and numbers. The manufacturing concept and procedures must match customer-driven elements such as responsiveness, flexibility, lead time, product dynamics, and life cycle.

Finally, the selected process characteristics and equipment must be examined for effects on manufacturing management, maintenance, safety, and the environment.

A plant charter/mission and an individual facility strategy will greatly aid facility planning efforts. There must be a significant financial analysis effort to justify the expenditure and to obtain formal approval and the commitment of funds (possibly from the existing capacity strategy set-aside). A facilities planning team must be organized; managers must be certain of which financial criteria will be used and the role of the new or modified facility.

The industry, its size and maturity, the typical increment of capacity and the capital required, and the balancing nature of such additions will affect how and at what corporate level the project will be managed. This can range from a corporate staff which both plans and controls (typical of process industries—e.g., ALCOA, petroleum, chemicals) to a division-driven plan and project management effort (focused facilities, short product life cycles, highly diversified and decentralized firms). Because of the usual magnitude of funds needed, most plans will be done one by one, in a logical sequence.

Depending on the industry and related product life cycle lengths, a facility may change both mission and charter a number of times in a 20-year life. Various models for change analysis are available, including those of *history* (what the firm or its competition has developed in the past, and what standards have worked well); *geography* (markets, transportation, economics, competitive locational advantages, total delivered costs considerations, "center-of-gravity" positioning); *Is it us?* (generic and specific functional needs and the ways these can be developed to match the corporate philosophy about facilities); and *specialization and focus* (competitor aggressiveness, the expectations of the customer, products, facilities and processes, and on manufacturing, for appropriate experiences to gain expertise in all aspects).

Coming Up (Next Chapter)

Facilities planning for manufacturing can be subdivided into the major areas of *location* and *design*. Location is strategic, and for best results is chosen in a strategic (representing) context by a multilevel team. Within design are the structure, services, layout, and materials-handling systems. The unique interactions among equipment and processes, layout and manufacturing concept, and the associated materials-handling system also call for teamwork. Today's facility plan is a good opportunity for simultaneous engineering through product and process design, as noted in Chapter 1. Overall facilities planning is detailed in Chapters 4 and 5. The location and design of the facility is an expression of the firm's strategy and is its face to the public. Locational analysis is not only a matter of dollars and cents, but involves marketing, logistics, availability of work force skills, the existence of educational resources, trends in the costs of the resources needed for production, environmental trends, and so forth.

LESSONS FOR THE INTEGRATED RESOURCES MANAGER

There are important relations among the 13 CIRM topics, and IFM stands at the center of many of them. In addition to those other elements of the manufacturing processes module, IFM needs to be involved with logistics, finance/accounting, quality, product design and development, human resources, the manufacturing concept, trends in regulation, and the community. The integrated resources manager must be an effective team player in both understanding and enabling the execution of an appropriate site development program. Individually and collectively, the facility *is* strategic and requires strategic attention. Because of the many interactions with all portions of the enterprise, the facilities study must be broadly based and given the resources needed for effectiveness. The integrated resources manager must take a truly *systemic* view of the role of the plant in the overall enterprise, before attempting to detail what the facility should look like and how it should perform.

Effective IFM must seek both design and operational guidance from a wide range of specialists within the company, as well as from external sources. This means teams drawn from the same diverse group of resources, to which must be added the necessary outside specialists in real estate, regulation and law, architecture, construction engineering, equipment design and development, and transportation. Such site development teams are vital contributors to the success of the facility in operation. IFM cannot be reactive and driven by a limited economic viewpoint, or the facility's life cycle effectiveness will be severely limited.

Chapter 4 provides some approaches to facilities planning, and Chapter 5 offers a practical demonstration of how the facility system can be analyzed, with specific examples of wastes and by-products.

APPENDIX

Nucor and Chaparral are steel suppliers, each of which has gained competitive position through the development of mini-mills, creating high-quality cast steel shapes from scrap steel, principally wrecked (salvaged) automobile bodies. Both raw material and processing costs are much lower than those of the integrated steel companies (see Fig. 3–8).

Steel products are created by a *process* facility—that is, batching is a matter of convenience, and the melting of steel scrap, its casting into bars, billets, or blooms, and the subsequent forming into marketable shapes frequently takes place continuously or nearly so. Approximate costs and labor hours data appear in Figure 3–8.

Experience in managing a higher level of technology can provide a competitive edge. Hayes and Wheelright call this a *dynamic* economy of scale. We also are aware of how first-to-market companies can both capture and exploit them and leave their competitors "looking at their tail lights."

FIGURE 3–8
Labor Hours per Ton: Nucor versus "Big Steel"

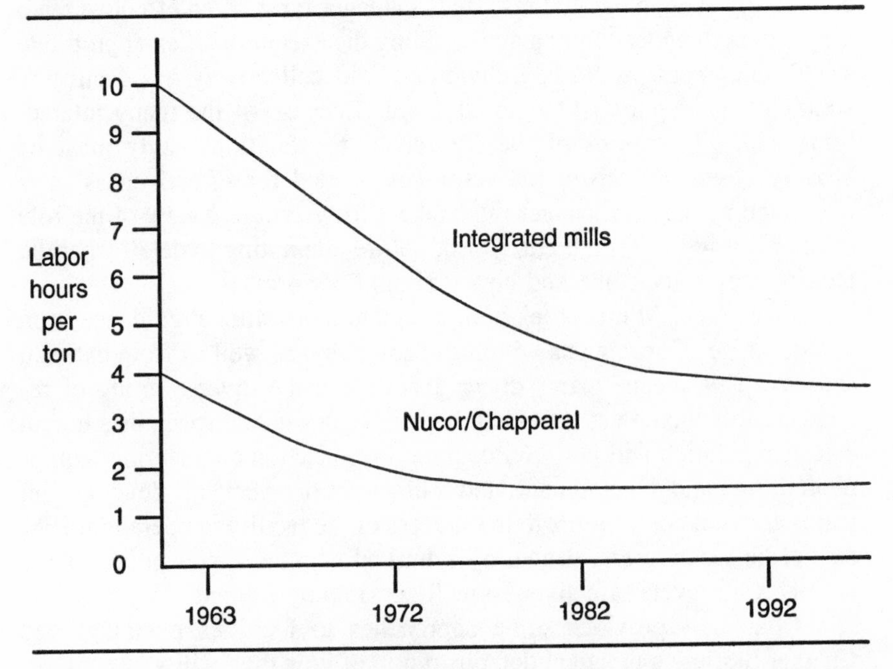

Source: John M. Burnham and R. "Nat" Natarajan, *Manufacturing Processes*, Student Guide (Falls Church, Va.: APICS, 1992). Reprinted with the permission of the publisher.

Nucor's long-term capacity strategy has been market driven (expand and locate to meet demand), where balances could be achieved in terms of customers, transportation cost, and scrap steel supply. Nucor was also process driven (learning how to manage more-complex shapes and metallurgies) Finally came the problem of adding value to the firm's products in terms of customer needs. Some expansion and forward integration was done through acquisition of other fabricator customers, where the plant's philosophy and strategy could improve operations and products. Nucor now has multiple facilities, each near a regional market.

The combined capacity and facility strategy of Nucor is strikingly similar to that of Chaparral Steel, a large mini-mill that has undertaken site development in phases at a single location in Midlothian, Texas. The most recent expansion enables Chaparral to do continuous casting of heavy structural shapes in almost finished dimension! Chaparral apparently feels that Texas offers a large enough market that economies of both scale and scope can be achieved in one spot.

The by-products and effluents of a mini-mill require green engineering to minimize environmental effects, making the mill a good corporate neighbor. (Further treatment of this issue appears in Chapter 5.)

REFERENCES

Bell, Robert R.; and John M. Burnham. *Managing Productivity and Change.* Cincinnati: South-Western Publishing, 1991.

Hayes, William H.; and Steven C. Wheelwright. *Restoring Our Competitive Edge: Competing Through Manufacturing*. New York: John Wiley & Sons, 1984.

Heizer, Jay; and Barry Render. *Production and Operations Management*. 3rd ed. Boston: Allyn and Bacon, 1991.

James, Robert; and Paul. Alcorn. *Facilities Planning*. Englewood Cliffs, NJ: Prentice-Hall, 1991.

Skinner, Wickham A. ''The Focus Factory.'' *Harvard Business Review, 10* (May/June 1974), 118–121.

Tompkins, James A.; and John A. White. *Facilities Planning*. New York: John Wiley & Sons, 1984.

SECTION 2

ENABLING

CHAPTER 4

AT THE INTERFACE: CONTINGENCY ANALYSIS AND PRELIMINARY FACILITIES PLANNING

Having decided on a greenfield solution, the team will immediately become concerned with alternative sites and their evaluation from the standpoints of accommodation of the preliminary design, workforce availability, access for construction equipment, and the raw materials and finished products to be involved. Formal or informal scoring systems will be used to rank the suitable sites, and good contingency planning becomes important to the chosen site's economic longevity. A lot *of bickering may take place.*

THE PRELIMINARY FACILITY PLAN

The next three chapters will move through the development of the physical production and distribution system as it successively affects *location choice,* initial and mature *site development,* and the various *physical facilities* that exist or will be constructed there.

The earlier stages of evaluation of these many general considerations—both externally and internally motivated—lead to the development of a preliminary facility plan, to be tested against various current and anticipated scenarios. As financial, market, technological, and specific site factors become known, contingency planning gives way to detail, and various architectural, construction, and equipment features are defined for bidding and regulatory approval purposes. Within the buildings themselves, office and plant equipment, structure, layout, and materials-handling systems are needed to carry out the site *mission.* This more detailed planning is presented in Chapter 6.

The preliminary facility plan leads to the production or distribution system component design, which, when examined across all the facilities in the system, reflects the strategy of the company. This is summarized in Figure 4–1.

What Creates the Need for a Facility Plan?

There are many different motivations for updating a facility plan: a new product, expansion for existing products, or the need to reduce capacity, among others. Market mix shifts can adversely affect a single site (particularly where transportation costs are a significant part of delivered cost), and major technology developments can force new study of both capacity and facility strategies. Green engineering addressing by-products and

FIGURE 4–1
Origins of the Facilities Plan

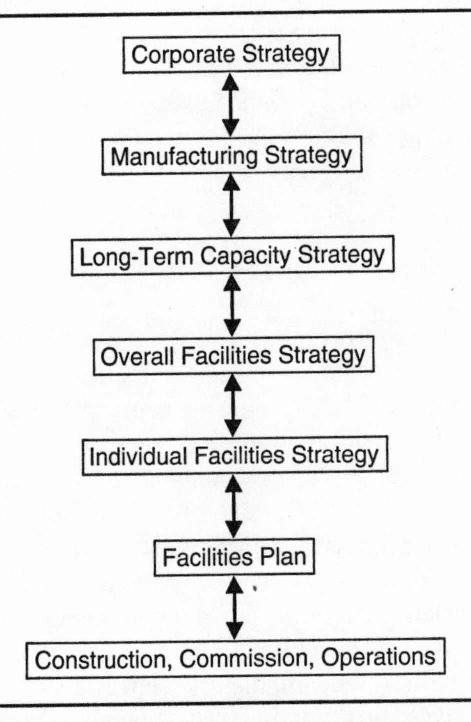

Source: John M. Burnham and R. "Nat" Natarajan, *Manufacturing Processes,* Instructor Guide (Falls Church Va.: APICS, 1992), Fig. 2–1e, pp. 2–7. Reprinted with the permission of the publisher.

waste can be the response to changes in either rule or enforcement timing of environmental or safety regulation. Finally, a facility plan can be developed to enhance customer focus and improve product competitiveness. A summary is shown in Figure 4–2.

Who Does the Facility Planning?

The industry, its size and maturity, the typical increment of capacity and the capital required, and the balancing nature of such additions will affect how and at what corporate level a project will be managed, ranging from a corporate staff that both plans and controls to a division- or plant-driven planning and project management effort (focused facilities, short product life cycles, highly diversified and decentralized firms). The term *site development* is often used to connote an ongoing program that starts with an existing set of conditions, develops the greenfield plan supporting the charter, carries it forward, and then maintains the site in operations and through successive grayfield modifications as strategic conditions dictate.

SITE DEVELOPMENT TEAM

To develop a truly effective facilities network, IFM requires a very broad view of the facility customers and of the suppliers of information to the planning process. Progressive companies are achieving significant benefits by formally empowering site task teams at many levels of the organization. The success of simultaneous engineering in developing better products more quickly can be shown, by analogy, to offer similar benefits to IFM efforts.

FIGURE 4–2
Triggers for Facilities Planning

Demand changes	Competitive benchmark shifts	Regulation
Market shifts	Competitive new product penetration	Internal strategic forces
New or unmet customer needs	Technology changes	Mission changes

Adapted from John M. Burnham and R. "Nat" Natarajan, *Manufacturing Processes,* Student Guide (Falls Church, Va.: APICS, 1992). Reprinted with the permission of the publisher.

A site development (SD) team should manage both the initial greenfield project and follow through over the economic life of the plant. Continuity over time and the inclusion of multilevel, multifunctional personnel are indicated. Such a team explicitly considers the many external and internal factors and influences discussed in this chapter. In this longer-term role, IFM is *representing* and working through strategic implications as an active team participant. The same team might carry this new knowledge forward into the detailed analysis and facility planning to be discussed in Chapters 5 and 6. In this intermediate-term role, IFM is *enabling* and addressing trade-offs and the tactics for implementation planning.

An effective SD team will be able to develop both horizontal linkages (with manufacturing, suppliers, process designers, and the distribution system or customers) and vertical linkages (the natural hierarchy), reaching relevant internal and external information sources to allow a detailed facility plan. Only under unusual circumstances could such a team be assembled from internal resources alone, with all of the necessary expertise to carry out the exploratory aspects of facility planning.

Consultants and Professional Services

By virtue of the many skills required to launch a greenfield facility or a major grayfield modification, a large number of professional people will be temporary SD team members. In some cases, they will be assigned from division or corporate staff, but quite frequently local knowledge will be mandatory.

Because of the manufacturing systems implications and regional marketing needs, much preliminary site selection analysis is done centrally. Real estate brokers/companies provide quiet acquisition services as well as local zoning and industrial development expertise. They work on commission, except for land assembly services, so selection is based on reputation and, perhaps, linkages through attorneys or prior company experience with the firm.

Competence of local construction contractors will be one of the determinants of location, unless a national firm is always used. Even then, much construction labor is locally hired. If architects are engaged to furnish construction supervision services, contractors must work closely with them. If competitive bidding is involved (rather than design-build negotiations), then formal solicitation, evaluation, recommendation, and award must take place, involving attorneys, sometimes assisted by consulting engineers/architects.

Architects/engineers are often part of the IFM/SD design team and may remain during construction and facility acceptance processes. They help with creative and structural aspects of the project and need to have access to all available strategic and operational planning information about the facility.

Financing institutions will certainly be involved, although some SD projects are funded internally or through corporate bonds. Many states or development districts are interested in attracting industry and will find ways to help with tax abatement, industrial revenue bonds, and other incentives, provided the financial arrangements are made locally. IFM will have a role, if locally financed, to carry out contracts made elsewhere. Divisional or corporate financial staff may become part of the SD team while these arrangements are finalized.

As land acquisition gives way to groundbreaking and construction, insurance underwriters, firms, and agencies may all be involved. Construction hazard insurance and workers' compensation coverage are normally the contractor's responsibility, but IFM may require bonds for liability and for completion guarantees. For fee purposes, local firms can provide local knowledge to assist with proper rating. Corporatewide coverages may already be in place and should not be disturbed. If the site has unique aspects, then separate underwriter investigations may be needed to protect all concerned. From the experience of fire and casualty insurance representatives, as well as the local fire department, IFM can gain greatly in terms of the various plans for site protection and for equipment and locations inside the facility.

Equipment suppliers are of great value to the whole SD team, but especially for manufacturing engineering, production, and maintenance specialists. Supplier engineers will help with installation and check out details and start-up problems, and they will give advice on spare parts and sourcing, and specific maintenance recommendations (not just "what the book says").

Location Analysis

The criteria for plant location, scale, and diversity are driven by the nature of the business. Company priorities may be influenced by markets, competitor locations, and shipping, manufacturing, and resource economics. This means that several sets of experts will be involved. According to Roger Schmenner (see References), product specifics and market area generally dominate the first location screening studies.

Real Estate

Significant second-level screening factors include land availability and cost; business taxes; ease of development; zoning; the quality, quantity, and cost of various utilities needed for manufacturing processes; and various state and local rules governing industrial use. Local specialists are needed to help evaluate site alternatives.

Input Resources

Both internal and external specialists may be involved to assess feasibility and economic advantage in regard to access to and the costs of critical resources: labor skills, costs, and availability, energy, raw materials, transport, and communications.

Site Acquisition and Development

Specific negotiating skills can help with the conditions, timing, and terms of purchase or lease. State and local industrial development boards can offer various incentives affecting both acquisition and operating costs. At the Nissan and Saturn manufacturing sites, for example, the state of Tennessee offered two major incentives: a six-lane spur to reach the nearest interstate highway, and substantial industrial training for the new production employees. These incentives amounted to $37 million (Nissan) and $52 million (Saturn), not including county and city agreements.

Environment

Systemic analysis must be accompanied by expert knowledge of both technology and legal requirements and their trends. Both state and federal regulations affect any industrial facility—the plant itself and the supporting infrastructure.

Corporate Standards

The Hewlett-Packard campus development process suggests the range of influence that may apply to consideration of any site. Appearance, size, and community relations may all be part of company policy.

Finance

Both corporate criteria and local financing options must be considered. The open and dynamic manufacturing system, when combined with ever shortening product life cycles, requires much more detail and contingency analysis to achieve LLCC for the initial and the mature facility. Estimating the timing of capital flows is important because the cost of capital reduces investment returns.

Internal Experts

A major benefit of the SD team is that specialists from product development and design, process design and manufacturing engineering, and facilities and manufacturing groups ultimately learn to work together and use a language that is mutually understood.

Location Advantage Examples

The logic of JIT manufacturing favors the location of design and manufacturing close to customers. New Jersey truck farms exist to serve the Metropolitan New York market.

Integrated steel mills tend to locate along public waterways, or where either iron ore or coal abound, raw material bulk and relatively low value make long-haul rail or truck transportation prohibitive.

Computer companies choose locations where skilled workers already are present and happy, and plan sites for both appearance and effectiveness.

On the other hand, Nucor and Chapparal placed their mini-mills where nonurban workers could be recruited—fairly near their markets, to save on transportation. They wanted to be near large population centers, however, so that they could obtain scrap automobiles and industrial and commercial construction would be in steady supply.

CONTINGENCY PLANNING

Because of uncertainty, some events cannot be explicitly factored into planning leading to contingency planning: exploring the unknowable and seeking logical action plans if one or another unexpected event comes to pass.

How do we evaluate the following issues?

New competitors or products, or both, or other aggressive competitor actions

Substitute industries, materials, or products (also competition)

Regulatory mandates (OSHA, EPA, zoning, labor, legal)

Unanticipated product successes or failures

Material shortages (war, transportation disruption, strikes)

New Competitors or Products

Marketing intelligence about customers and competitors is gathered continuously. This kind of competitor environmental tracking is a specific responsibility for business strategists at the corporate level, but frequently this intelligence is not passed along to the manufacturing system until there is only time to react, not to plan. For example, output levels, mix, and the timing of production may all be affected by a significant competitor price change. IFM teams must seek to stimulate the flow of relevant guesstimates, as well as hard data, to manage the required changes.

Substitute Industries, Materials, or Products

Sadly, the impacts of such substitutions are not recognized by the various current industry participants. With a vested interest in steel-forming technology, the Big Three auto assemblers and their traditional suppliers were slow to recognize the impact of rising fuel prices on car weight. The U.S. electronics industry was slow to adopt solid-state devices into their consumer products, allowing Pacific rim countries to gain a dominant market share and eventually drive most US firms out of business. An industry that is first experiencing such substitutions is particularly vulnerable to their impact. The recent disclosure of IBM's difficulties show the problems of a rapidly changing market and a slowly adapting product positioning strategy.

IFM teams need to maintain good communications lines with R&D and engineering technologists, both within the company and through professional organizations, in order to help manage capacity and facility impacts that can otherwise devastate the company.

Regulatory Mandates

IFM at the plant site carries a specific responsibility to track regulatory developments and to keep communications open with local and regional regulators, as well as corporate monitors and lobbyists. Knowledge of trends can prevent surprises, especially since public hearings accompany major change proposals.

Unanticipated Product Successes or Failures

Keeping good ties with sales and marketing personnel (networking) is important for dealing with changes in demand. From the IFM standpoint, the focus facility with multiple machines is much easier to manage in either expansion or contraction mode. This is especially true if machinery is "standard" or can be customized by the plant shops. The appropriate manufacturing concept (focus, cells, SDWT, flow) can make any product or capacity transition much easier. Sometimes the success or failure is not directly related to the individual facility's mission, but to the plant role as a supplier to or customer of another plant whose production schedule has been suddenly impacted.

Material Shortages

Oil embargoes are not the only instance of recent material shortage, but the whole range of petroleum products and their derivatives offers a good example of how material shortages can damage an industry. Especially with JIT, even short-term transport disruptions can have serious effects. The impact on Saturn's assembly activity of a strike at GM's Lordstown parts plant comes immediately to mind. IFM's involvement may be with substitute materials or processes or with layout realignments to make rather than buy the shortage items. These problems are certain to arise with production in a global market, where overseas sourcing and overseas markets are the usual form of doing business.

General Approaches to Contingency-Based Facility Planning

From the SD and IFM standpoint, the goal of contingency planning is to determine the proper mix of initial facilities, standard expansion increments, and inherently easy-to-modify services and utilities to accommodate the unknown.

Many challenges have arisen within the automotive and aerospace/ defense industries. For automotive manufacturers worldwide, the challenge has been to meet or exceed rapidly changing customer expectations about quality and performance, governmental interest in fuel economy and emissions control, and vigorous worldwide competition and market segmentation. The impact on facilities has been very great, as modernization and expansion have taken place concurrently with plant closings and layoffs. Most domestic automakers remain under siege.

For the global aerospace and defense industry, the "peace dividend" following the fall of communism has had serious repercussions. Textron Aerostructures in Nashville, Tennessee, has over the years moved from nearly 100 percent U.S. military component contracts, through diversification (domestic motor coach construction) to its current substantial commercial contracts in worldwide markets. Customer demands are still continuing to change how Textron manages technology, planning and control processes, needed personnel skills and numbers, and ways to organize work.

Plant Changes for Customers and Competition

External factors maintain pressure on the entire manufacturing system, but with ultimate focus on the individual plant. The purpose of a plant mission or charter is to define the ways in which the firm's strategy will be implemented throughout manufacturing. The manufacturing concept establishes the desired relationships among all the factors of production. Both the details and the charter are dynamic. When customer expectations change (about value, quality, or timeliness) or when world-class competitive benchmark shifts are unfavorable to the plant, adjustments will be needed to maintain a competitive posture. Design features reflecting the results of contingency planning and what-if analysis can be of two kinds: modularity, so that expansion can be accomplished without disrupting the existing operations, and flexibility, so that incremental internal plant reorganization can take place without major teardowns.

Simulation

Complex situations may demand a computer simulation of the required systemic materials movement and evaluation of alternatives to meet the needs. These simulations have been applied to pharmaceutical plants, paper products companies, and integrated rolling mills like ALCOA. Scale mockups are used for interference modeling and for personnel training.

The decision to simulate or mock up can be more than justified by the time and investment funds saved. In plants where the decision to automate materials handling has been made without evaluating the system for value added, the result may be automatic guided-vehicle systems (AGVS) for moving parts from work center to work center, massive conveyor systems to avoid the tow motor (and operator) expense but losing inherent flexibility, and needing complex computer control to manage the movements.

CONVERTING UNCERTAINTIES INTO CONTINGENCY PLANS

The example that follows includes some details about how an organization can be proactive in dealing with the future.

The electronic components, test equipment, and computer industry is very dynamic. New products have their day and rather quickly trade on price alone, having been supplanted by superior attributes in newer products. Competition is intense and unrelenting, and share of market is *never* guaranteed. How, then, does a company make plans for research, manufacturing, and distribution facilities?

Hewlett-Packard and the Electronics Company Campus

Hewlett-Packard (HP) is global in every sense, having facilities on five continents. It is a leader in implementing JIT/TQC and has moved to global logistics management as well. It sells its products in a wide range of markets, its margins are good, and customer service levels are high. From the capacity standpoint, HP cannot afford to fail in meeting demand.

As will be recalled from Chapter 3, there already exists an overall facilities plan for each site (called a *campus* by HP) and a tentative timetable for its development. The initial campus structure is divisional headquarters, which houses R&D, engineering, marketing, and other staff functions. Initial new-product manufacturing takes place there as well, with appropriate staff support. As demand for a new-product family increases and product development bugs are worked out, successive structures are erected to house assembly and test for the present and future product families.

What lessons can be learned here? Land acquisition should allow for the ultimate size of facility construction and operation, and the entire campus preliminary plan is known. Only a single building will be erected at first, while the new product "division" seeks to find its market. This reduces up-front fixed capital requirements and gives better matching of added costs and revenue growth.

Considerable time may elapse before the product family (or market segment) is sufficiently established to warrant expansion. Because of the global HP presence, components (switches, power supplies, PCB/modules, motherboards) will be purchased and brought to the campus for assembly, rather than going into full-scale component production while demand is uncertain.

What happens as product demand matures and declines? New-product development is still taking place at the parent facility and will be transferred to production when demand has been established. As demand increases, justifying more expansion, the rate of building growth on the campus increases, more assembly capacity follows, and a component plant is created to serve the booming products demand.

When current production capacity exceeds demand, reversal can take place. Excess demand from other campuses can be fulfilled by this site, so long as the process and assembly technology matches the "sub-contracted" product needs. The campus can sustain operations until obsolescence overtakes it. Then new teams, research, development, and prototyping can come in, supporting new equipment and tooling, and the rejuvenated site can return to leadership once again.

WHAT PROCESS SHOULD BE FOLLOWED?

The result of detailed facility planning is a set of design plans and specifications and, frequently, an action plan—a timetable and sequence in which various component tasks will be carried out. The plan is driven by the site mission or charter and must reflect everything currently known about products, processes, and required volumes. Such a set of documents is developed iteratively and with considerable overlap among representing and enabling activities and the responsible SD team(s). What follows is an outline of how the design evolves.

Assessment

To carry out the analytical steps included in most consulting, engineering investigation, or other problem-solving procedures, it must first be determined what *is*. Such an assessment examines the existing plant for its product assignments and the current and expected volumes for each product or group. These are then compared with current and projected capacity. For a greenfield plant, the charter becomes the performance specification guiding design.

Requirements

Mandated plant capabilities are then examined, especially those related to quality and flexibility. Plant organization generally is determined by the manufacturing concept as well as the product assignments, so focus by either product line or process may be important in understanding plant operations. The sum of the foregoing capabilities in large part determines cost—another plant "capability" that is an important part of the assessment database.

Comparisons

When these descriptive aspects are then related to the plant charter and mission, mandated changes in capacity or capability become very visible. The nature and extent of changes make up one leg of the plan; the timing for the capacity and capability additions makes up the other.

Study

With a new facility's design features set out, a feasibility study by the site development group (or perhaps the multilevel corporate and divisional team) must consider the range of available alternatives through which to get the needed result. A variety of financial and competitively determined technical and operational constraints must be considered before a plan is developed.

Outlook

Since the facility life cycle is considerably longer than that of most product lives, the probable future requirements must be examined as well. Through *contingency planning,* many possible future conditions are evaluated for their effects on the site, to discover what provisions can best be made to minimize negative impacts over the plant's life. The objectives of such what-if planning allow for the greatest effective capacity of the proper kind, so that the lowest life cycle costs will be achieved.

Tying the results of contingency planning into the facility design is truly a matter for teamwork. And the diversified team is the most effective means of designing for the LLCC, although chaos increases! The result is a recommendation to act: setting forth the plan, supporting logic, timetable for accomplishment, and resource requirements.

This iterative process—description, comparison of needs with capabilities, evaluation of means for accomplishing results, contingency planning to address the uncertainties surrounding the facility over its lifetime, and settling on the best LLCC design—is comparable in every way to the product development process and, done properly, carries the same benefits forward into plant operation.

Maintenance Planning

IFM has the responsibility to develop the plant's maintenance program to provide equipment with sustainable *quality capability*. The maintenance planning effort—at the enabling, or tactical, level—is close-coupled with manufacturing and scheduling. It also is coordinated with quality personnel, because their analysis of tolerences and variability begins in the process design stage and continues throughout the life of the plant. Because of its critical importance to all aspects of cost, operations, and customer satisfaction, Chapter 8 is devoted to maintenance.

Facility Plan Evaluation Criteria

Once the needs and forecasts have been detailed and the policies governing design solutions set forth, what-if analyses can yield an optimal initial solution and minimize the cost and production impact of later grayfield modifications, assuring the lowest life cycle cost for the site.

LLCC is the primary target of IFM and the site development team. By definition, LLCC is the discounted sum of all initial and estimated future costs of capital, maintenance, and modification programs over the life of the facility, taking into account company policies, plant mission and charter, and estimates for relevant factors whose values are uncertain.

Life Cycle Considerations: Optimizing Business Effectiveness

A number of uncertain factors make "optimality" hard to prove. But careful and objective assessment and the procedures outlined above provide the best trade-offs between design detail and contingency allowances. That is, the initial design should reflect all that is known, and allow, through flexibility and adaptability, for what is not known. As noted in Chapter 3, some of the design aspects are long-lived and become part of the site's "fixed capital," available to support any manufacturing assignment. Others are product related and are only expected to serve during a specific life cycle.

Fixed Capital

Building Structure and Arrangement

The preliminary facility planning addresses those aspects that are best designed in. Most of the following examples become part of plant fixed capital, rather than being costed against the first products assigned to the location:

- Major utility lines (e.g., the water supply and drains for a textile-dyeing operation) need considerable what-if planning to avoid expensive downtime later.
- If later product assignments will require intallation of much heavier equipment than the start-up plant requires, then proper foundations in the general area would best be fabricated during the initial greenfield work.
- If heavy materials-handling equipment will be installed, determining how and where it may be needed over the life of the facility can lead to relatively low-cost greenfield strengthening and roof height planning. A drastic alternative could be abandonment of the building and large write-offs.
- There are many building perimeter penetrations—requirements for natural light, forced ventilation, and personnel access, for example. Others are associated with production functions, especially for materials movement.
- Loading and receiving docks may be at opposite ends of the plant because of the production flow logic. But JIT may later seek point-of-use delivery and storage, with transport vehicle access along the entire length of a production/assembly line. Although installing a large door instead of a blank wall seems simple enough, it is only quick and inexpensive if the structural supports of both walls and roof are spaced to accommodate this future need!

The approach suggested in each of these examples may add to the initial project cost, but it can greatly reduce future modification costs.

Considerable effort accompanies major facility revision, perhaps requiring that SD teams reduce some operations support activities to carry out the grayfield planning. And there are the opportunity costs of later grayfield downtime, through lost plant production capacity and idle crews.

Allocable and Product-Related Costs

Specific product knowledge allows the SD team to examine such issues as life cycle and sales volume patterns. Taken together with the manufacturing process, these can affect equipment choices and layout, structure and foundations, and various utilities that support production. Short life cycles can mean frequent changes. SD plans must address potentially affected areas, choosing carefully what should be capitalized, what should be deferred until needed, and what should be expensed as product-specific.

Equipment LLCC

The length of the product life cycle may suggest that only machinery and fixturing with equivalent economic life should be used, but some major equipment may be part of a fundamental process step (e.g., ALCOA's hot line, or HP's IC wafer preparation equipment) and become part of the site's fixed capital to be recovered over many different products using the same technology.

Site/Community LLCC

Most of the provisions that establish and maintain good relationships with the community are fixed capital. Attractive and durable exterior and interior design, neat and easily maintained grounds, parking areas and access routes that are fresh and easily traveled, visitor spaces near the business offices, good lighting throughout the grounds, and so forth, all present a face to the community and to the employees that projects what the company believes is important. Unobtrusive but efficient materials transport routes, waste disposal areas, utilities tie-ins, cooling towers, and screened outside storage can also contribute to the site and company image.

Interior amenities in terms of work areas, personal lockers, restrooms, and cafeterias used to be the hallmark of relatively few companies. But with the rediscovery of untapped employee potential, much more future IFM effort will be directed toward attractive as well as functional work space. In addition, team concepts are being applied broadly across what used to be separate office and production personnel and areas.

Regulatory Compliance LLCC
Taken broadly, this can involve products and processes, as well as the facility itself. Good greenfield design will not only study current safety and environmental regulation but will reflect the company's concern for employees, community, and its own reputation as a good corporate citizen. There are always choices as to energy sources, process specifics, and the means selected to handle by-products and wastes. There are similar choices in simply complying with, rather than anticipating, future standards for air, water, safety, and so on.

Good SD teamwork can lead to better industrial "ecosystems" and to lower LLCCs as well, because most retrofit work (locking the barn door after the horse is stolen) is only cost-added and not value-added. Anticipating regulation and working to maintain full compliance over the life of the facilities at the site is one of today's larger IFM challenges.

EXAMPLES OF CUSTOMIZED FACILITIES

Capital Intensiveness

Chemical plants, refineries, and electric power utilities—being process plants—and large-scale consumer-oriented facilities like the Procter and Gamble Paper Products complex at Cape Girardeau, Missouri, also fall into the customized category of planning and implementation work. The general characteristics are the same:

A major investment

A long recovery period or a high-margin, differentiated product

A large increment of capacity

Unique technical and economic features

Permanent structure to accommodate major equipment

Need to project both future demand and the nature of future services

Mostly irreversible investments due to uniqueness and no ready resale

Mandate for practical optimality in meeting *all* corporate goals

As a matter of course, this kind of bet-the-company project leads to highly centralized planning and control. The project management team involved in the contruction phase will be made up of technically experienced,enced, on-site people, whether from the company or from an architectural firm. This team will work closely with corporate staff, especially at various critical junctures in the outfitting and testing phases, when the paper design will be tested against the reality of the plant as built.

The headquarters building is formally recognized to be an expression of "what we are," while operating facilities may not receive the same level of attention. Nucor and Lincoln Electric are challenging exceptions to the luxurious structures erected by some companies. Offices are spartan, with highly functional design and little excess space. In one sense, the offices at Nucor and Lincoln *do* express "what we are" because they represent the corporate commitment to lead the pack in terms of high quality and low cost. Significant investment in plush offices is simply inconsistent—a message understood by both customers and employees.

Modular or Multipurpose Plant

At the other end of the facilities spectrum is the standard Stran or Butler building structure that is erected on a concrete slab, using prefabricated beams and panels, and adaptable to a variety of tenants. Only minimal specifications are needed: enclosed square footage, floor loads and penetrations, floor-to-beam clearances, loading- and receiving-dock numbers and configurations, and office spaces external to the plant. The contractor can do the rest. Here, the nature of the task is straightforward and involves location and site orientation, construction supervision, and outfitting once the building can be occupied. These planning characteristics include

- A modest investment
- A short recovery period
- A small increment of capacity
- Standard technical and economic features
- Adaptability to changes in demand and the nature of future services
- Flexibility of equipment and layout, readily modified or relocated
- Little or no uniqueness—easily modified for another occupant
- Fit for resale

In this modular mode, IFM tends to focus more on the details of support for the manufacturing customer—structure, layout, and materials handling—and for the human and regulatory participants, through safety, environment, and operational utilities support. Note that with proper site size and zoning, similar buildings can be erected and outfitted to handle increased demand with little or no interference to the existing operations.

Company Standards

It is also important to understand what criteria will be applied to site development proposals and what restrictions or policies may apply to scale, design, and contracting procedures.

ROI Mandates and Competitive Positioning Needs

These elements are sometimes conflicting, especially when profits are being squeezed by market share loss or competitor price cutting. A related issue is the time frame for cost justification. In any case, these data must be integrated with the facilities plan. If a facilities budget is dependent on attaining a specific "hurdle rate of return," then equipment choices and operating practices may be restricted.

Company Facility Policies and Criteria

A number of companies have standards for buildings, site development, and capacity increments. Offices and amenities, company cafeterias, break rooms, restrooms, and security may have policy guidelines simplifying the design task. Sales offices and distribution centers may also be subject to company design standards. Depending on the nature of the business, some companies support a corporate facilities group and actually carry out major facilities programs under centralized direction. As noted, process businesses tend to have this kind of organization.

Equipment Standards

Communications, data management, and fabrication activities often need to achieve commonality for critical components and for software or control systems. Given the increasing degree of interdependence, there may be policies or standards in other functional areas that can affect site development. Industrial relations, environmental protection, safety, policies on building, equipment, liability, and disaster protection are all likely subjects for corporatewide standards affecting individual company locations. As mentioned earlier, Hewlett-Packard has evolved a site development program governing location, size, appearance, and individual structures, as well as the economic or market criteria to determine when any expansion will take place. HP also has global standards for sales and administrative offices.

Thus, standards can exist for total site development, along with construction and industrial standards to "size" power, ventilation, lighting, passageways, insulation, and so forth. The use of these general and technical standards can help make the IFM task more manageable and allow focus on special customer needs.

The Charter

As already noted in Chapter 3, a plant charter/mission and an individual facility strategy will greatly aid facility planning efforts (see Figure 4–3). Managers must be certain which financial criteria will be used, as well as the role of the new or modified facility. The charter and associated plant mission serve to relate these criteria and are supported by the capacity and facilities strategy position. The preliminary facility plan is used to justify the expenditure, seek and obtain formal approval and the commitment of funds, and undertake the organization of a facilities planning team to carry the work forward.

The plant mission, like the company in its marketplace, is dynamic and subject to changes. Strategic shifts in company positioning can lead to redefinition of the charter assignments for part or all of the facility network. Decisions to introduce new products or terminate mature ones lead to changes in the charter and in the relationships among facilities. Production facilities can take on distribution center responsibilities and vice versa. Growth of product demand can require added equipment capacity. A shift in consumer preferences can alter the mix of options and related manufacturing volumes. Strong growth can force secondary products out of the plant entirely, leaving a de facto dedicated facility. But a dedicated facility is also vulnerable to demand downturns, which may lead to proposals to bring other buildable products into the plant charter to maintain employment and production volumes.

LOCATION FACTORS

The following contingency questions and responses focus on the location decision and can serve as an assessment tool for site selection.

FIGURE 4–3
A Plant Charter

Products and volumes (by item or product group)
  Volume capabilities
  Rank order of priorities required
What must the plant do well? (mission statement)
  Key leverage points and critical competitive factors
  General scenario for expansion/development (facility life cycle)
Process capabilities and capacities
  Capabilities by type of production process
  Capacities
  Changes over time
Development directions
  Plant and equipment
  Production planning and control
  Labor and staffing
  Engineering
  Organization and management
Specific objectives (by product and year)
  Unit costs
  Service level
Capacity utilization
  Scrap loss
Inventory investment
  Cost structure

Source: William H. Hayes and Steven C. Wheelwright, *Restoring Our Competitive Edge: Competing Through Manufacturing* (New York: John Wiley & Sons, 1984). Table 4–4, p. 100. Used with the permission of the publisher.

1. How much current (versus future) capacity is needed, and *where* in the manufacturing system should it be located? The clear trend, evident for the past 20 years or so, is toward smaller, regionalized, focused plants (Schmenner, 1982). Rather than following the historical practice of a few large plants with separate regional distribution centers, these smaller facilities are often combined with distribution. Thus, the *overall* scale of production facilities will continue to shrink, gaining flexibility and ease of management, but sacrificing some economy of scale. Because the overall capital expenditures will be divided into smaller and more evenly time-phased increments, this reduces financial risk for the company, as capacity can stay reasonably close to that desired for maintaining share of market.

2. What are the trends in real estate: availability, zoning, incentives, prices? The glut of commercial real estate and office-building development triggering the S&L and banking crises has significantly brought down both land and building costs. Given the resurgence in population growth and the steady pressure of immigration, this glut is only temporary. Other factors will keep the task of site selection a challenge. Note the role for the real estate specialist in helping to pin down and quantify these trends.

Locational analysis is driven by many criteria, including customer density and transportation. Other necessary elements of the information gathering and analysis stage of site selection include how much land is needed, what manufacturing processes will be permitted, what environmental factors apply, what waste disposal problems might occur, what local or regional economic factors might lead to important financial concessions by the locality (tax abatement, industrial bond financing, state-funded training, roadway construction, employment recruiting support, and so forth), and finally, what price and financial terms apply to the land.

3. What are the locational needs: access to raw materials, work force skills, "presence" in a community, and access to customers? More traditional locational analysis can take place concurrent with the macro factors noted above. Logically, some plants *must* be reasonably near the source of raw materials (the mine, forest, or oilfield) or the customer (newspaper printer, container fabricator, insulation manufacturer). But equally compelling may be the availability of the needed number of workers with necessary skills, the attractiveness of the location for the professional people that the plant requires, or the "political" or marketing importance of having a production or distribution facility located in the region.

The critical issue today is the ability to meet customer expectations. JIT calls for frequent, small-lot deliveries, which means either a nearby location or excellent, low-cost transportation to reach the industrial customer. Consumer goods manufacturing may need to be located near the customer, but inventory costs must be minimized. Global markets demand global logistics management, integral with manufacturing and distribution center development. The choice of many smaller facilities allows considerably greater flexibility, especially when office, distribution, and production functions are combined at a single site. These factors once again demonstrate the need for *integrative* facilities management.

4. What must be achieved to give the company/site competitive advantage: flexibility, target lead times, production and transport

costs, access to key suppliers or customers, regional capacity dominance? This translates into the plant charter/mission, congruent with what is needed to support the manufacturing strategy and the marketing/product strategies. The site development program must reflect the practical optimum *over time* that balances all these factors to give the company the greatest advantage possible (*not* necessarily the site when evaluated as an independent entity).

5. What form should the capacity take? Should equipment be special-purpose or general? What capacity choices are available, given the needed scale of process(es) and the manufacturing concept? Economies of scale in the plant overall or in specific equipment can be lost when the products are dynamic and high-volume equipment is no longer needed.

This happened to Black and Decker, which had to close the flagship plant in Hampstead, Maryland, because it was designed for high volume and low variable cost, but changeovers were extremely time consuming. A different situation affected GM Hamtramck and John Deere: large investments in highly automated, numerically controlled equipment did not yield the expected returns because the basic manufacturing concepts did not fit the new equipment's capabilities.

Other issues include how manufacturing will be organized (group technology/cell manufacturing, U-lines, product lines with little diversity, one-product plants with variations) and what will be required of people. Decisions about these factors will affect site selection and development, structure, layout, and so forth.

6. What is known about the product dynamics: life cycle, rate of change, diversity, complexity, material component substitutions, lead times to customer? From the site development standpoint, the mature product (but with staying power) is much easier to plan for than a new, hot item with no track record. Paper plants, steel mini-mills, and petroleum refineries are in this mature category. From the design standpoint, mills and refineries do not change a lot after construction. Efficiencies are critical, and capital investments can be recovered over a long time. Generally, these process industries will choose larger facilities, since the "six-tenths rule" allows *doubling* the design capacity of a new plant for only 60 percent additional capital. Equipment selection will be toward the big machine ($50–$250 million) or the high-capacity refinery.

On the other hand, electronics component and computer makers find themselves in the hot-item, short life cycle domain. When new construction is planned, there is neither a way of predicting just what products nor a way to predict what processes will be at the facility over its 20 or more years of economic life. This will lead to very different choices for plant size, equipment, fixturing, tooling, and materials handling. Life cycle costing will be based on the initial product assignments and their projected length of run, leading to the goal of optimizing the product-specific equipment and facility support to recover those costs within the product lifetime. More general-purpose equipment can justify a longer recovery period.

For the hot-item producer, significant facility modification will be almost continuous, and site design will seek to minimize the ongoing development costs. Modularity, adaptability, and flexibility must have very high priority in facility planning here.

7. What is known about process dynamics: fundamental process obsolescence, equipment lead times, and maintenance and operations workforce skills? The technologies for manufacturing, as would be expected, are highly responsive to the needs of the customer—in this case, the IFM site development team. Large capital investments require long analysis lead times and will probably call for all the "bells and whistles" of the most up-to-date systems available. Therefore, large capital equipment developers and suppliers strive to be on the leading edge and to help the site development team achieve the long-term efficiencies needed but with proven technology. The fundamental paper-making (Fourdinier) process did not change for 175 years after its invention, for example. The same is true of steel and nonferrous metal processes, oil refining, and textile manufacturing. Although dynamic today, these changes seem somewhat more gradual and predictable. Costs can be absorbed over the long productive life of the facility and charged as overhead to product.

But even in traditional fabrication and assembly businesses, mechanical processes are undergoing change. Greater range of diversity, more precision (quality), electronic controls, and greater adaptability and flexibility are now customer requirements as well as a justifiable life cycle cost. And fundamental materials changes have required major process revisions. The shift from sheet and cast metal for housings and cases toward high-impact plastic has caused almost as great a revolution as the move from vacuum tube to solid-state electronic technology. To stay competitive, existing equipment is obsoleted and sold off almost at once, and new equipment and new technology must be acquired. New suppliers must be developed or existing suppliers retrained. In addition, waste disposal problems are multiplied.

In the solid-state electronics field, the speed and size barriers have been tumbling for the past 40 years, with enormous impact on facilities and on *their* customers: R&D, engineering, manufacturing, and the ultimate purchaser. Designers, engineers, and marketers are all present on the factory floor with production people to make sure that the product meets customer requirements, includes the best design, and can be manufactured with low cost and high yield. After all, with the up-front cost of a new microprocessor chip well over $100 million, being first to market with a superior product is the *only* winning strategy. The issues here are very complex for chipmakers like Intel, because the capital equipment for very large scale integrated (VLSI) chips is very expensive and each generation of chips is more closely packed and faster than the last. Here, manufacturing technology develops hand in hand with the design work and with the determination of customer needs.

In terms of investment analysis and life cycle costing, how is this handled by electronics firms? Product-specific items (including design, masks, and special tooling) are charged directly to product and must be justified on the basis of expected product life. Generic investments (chip slicers, laser imaging devices, clean rooms, and so forth) can be depreciated over the life of the *technology,* not the product, but still not over the life of the facility, as was traditional. This so-called fixed capital invested in the facility is an ever *decreasing* portion of the total and is therefore much less significant than for stable products and processes. Chip consumers like Hewlett-Packard take this as a base and add all other costs (labor, materials, tooling) into a conversion cost pool for ultimate charge-off against products shipped.

8. What are the trends in the societal environment: local zoning and effluent issues, employee health and safety regulations and trends, liability trends and insurance coverages? Major challenges to site development lie in this "societal" domain. Both public and private groups are increasingly active on behalf of the environment. What was accepted practice a few decades ago has led to plant abandonment: complete shutdown because of the economic inability to come into compliance with court mandates or regulations. In some respects, this area is uncertain, but the trends are quite clear. Environmental-impact studies are mandated before major industrial (or residential!) construction will be permitted. The same principles of systems analysis that serve engineers so well must be applied to product, process, and facility, but with the broadest possible definition of what the system is. All inputs to the process, all outputs from the process, and careful characterization of the effects of both on people, water, air, flora and fauna, and longer-term effects on ozone, polar ice caps, and distant vegetation must take place routinely. Chapter 5 adds both analytic approaches and examples of good by-product management.

9. How will just-in-time, total quality management, and total productive maintenance be reflected in physical organization (layout, support staff locations, meeting areas, training rooms) and in the plant layout dynamics? Equipment layout will certainly change, and quality will be integrated with production. Materials-handling systems will reflect one-at-a-time or small-lot manufacture. Formal or informal self-directed work teams (SDWT) and former "office" personnel colocated into production spaces call for desks, meeting areas, access to data, and places and equipment to conduct experiments. These mixed facilities will affect air conditioning and ventilation standards, noise levels, airborne vapor, forklifts and other materials-handling gear, safety, emergency planning, and so on.

LESSONS FOR THE INTEGRATED RESOURCES MANAGER

In this chapter, the varied elements of what makes up the manufacturing *system* have become more tangible through a focus on the facility within which actual production will take place. Proactive plant design is hardly a simple process because so many factors, both internal and external to the company, must be considered. Figure 3–2 expressed this variety and suggested some important relationships.

There is considerable corporate-level strategic involvement among many functional groups as the SD team's representing role is fulfilled. Capacity, plant size and plant groupings, and supply and distribution are all involved before internal-to-plant specifics can be analyzed and designed. All of these are influenced by the customer for the products to be made and by the firm's performance criteria for the facility. Technically, the plant structure, equipment layout, and materials-handling systems are all affected by the ways that people and equipment are integrated for effective operations. A list of specifics is less persuasive than adding some detail to the CIRM diagram shown as Figure 4–4.

FIGURE 4–4
Additional Internal Relationships Involving IFM

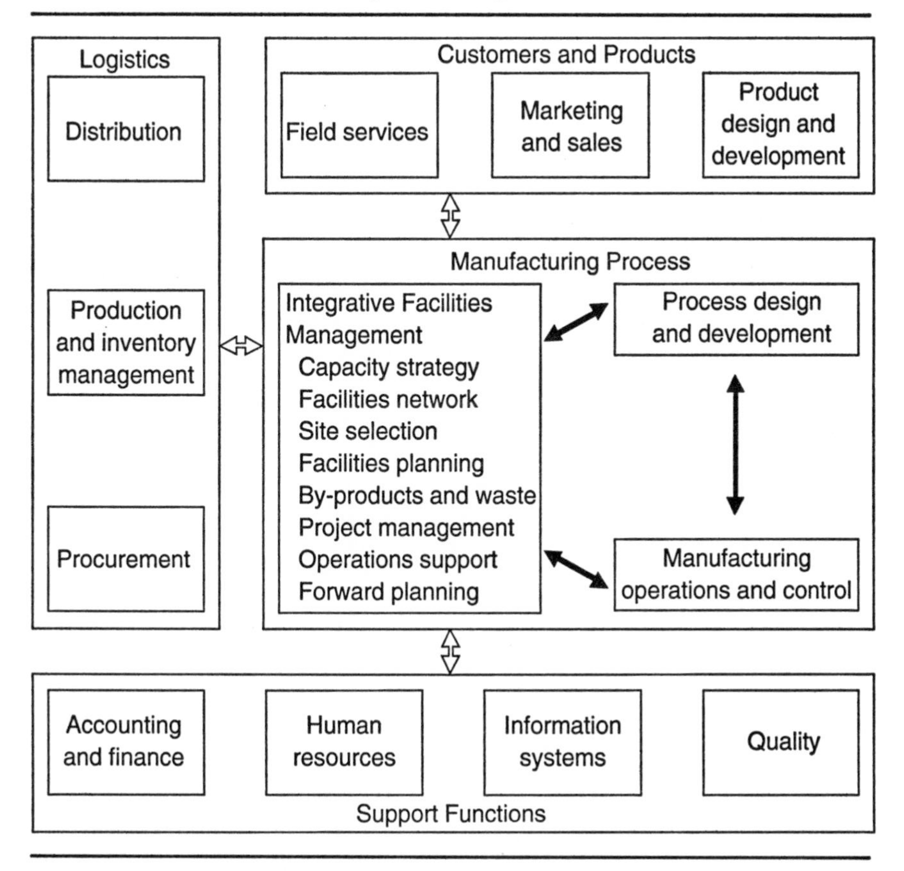

Teams

The formation of a site development team was emphasised as an effective means of achieving the many IFM assignments. The benefits will be substantial, for 35 to 50 percent of total industrial company assets are invested in various facilities. There are many tactical improvement opportunities in equipment selection, structure, and machinery selection, but the greatest benefits will come from the clear focus on what the facilities must contribute to the company's global competitive advantage. It takes management initiative to move facilities management practices from reactive to proactive and to incorporate the inputs of many other external and internal disciplines into the facility planning process from its inception.

Change

External factors can cause a new or modified plant mission or charter guidelines, redefining various aspects of the manufacturing capabilities required by the marketplace. The modern plant must have an appropriate degree of adaptability. Expansion of capacity, change of product mix, addition or deletion of products, conformity to new quality demands, and even plant reconfiguration will be frequent incidents in the lives of most facilities. Regulation can have many effects on production, by-products, and performance criteria, serving either to mandate action or to prevent it.

Self-Generated Change

In addition to market-driven or regulatory changes, many others are internal and, perhaps, voluntary—for example, deciding to move toward flow manufacturing, machine cells for parts families, or strong employee involvement, cross-training, and self-direction programs. These changes will trigger frequent physical and procedural modifications to the goal of continuous improvement.

Cost Impacts

Almost any change increases plant costs (although one would hope that many would have revenue benefits too). Part of the continuing responsibility of the SD team is to track these changes, and alternative ways to address them, throughout the design process. The target is lowest life cycle cost because that offers the greatest profit opportunity for the products and services being provided. The "accounting" for LLCC is complicated by another dynamic, that of shortening product life cycles and calling for shorter redesign times. This leads to the separation of facilities costs into two categories: fixed capital, without which the facility could not fulfill production responsibilities for *any* company product; and product capital, those machines, tools, and fixtures that can be directly tied into a specific product or family. The former is handled like depreciation and is charged back over the life of the facility. Product-related capital must be recovered over the life of the specific product or family. An effective SD team will evaluate the various alternatives for buildings, processes, and machinery from both cost and operational viewpoints, because operating and maintenance costs are also a part of LLCC. The great range of significant interdependencies is basic to the development of the CIRM program and this text series.

REFERENCES

Bell, Robert R.; and John M. Burnham. *Managing Productivity and Change*. Cincinnati: South-Western Publishing, 1991.

Hayes, William H.; and Steven C. Wheelwright. *Restoring Our Industrial Competitiveness: Competing Through Manufacturing*. New York: John Wiley & Sons, 1984.

Heizer, Jay; and Barry Render. *Production and Operations Management*. 3rd ed. Boston: Allyn & Bacon, 1991.

James, Robert W.; and Paul A. Alcorn. *A Guide to Facilities Planning*. Englewood Cliffs, N.J.: Prentice-Hall, 1991.

Schmenner, Roger W., *Making Business Location Decisions*. Englewood Cliffs, N.J.: Prentice-Hall, 1982.

Skinner, Wickham A. "The Focus Factory." *Harvard Business Review, 10* (May-June 1974), pp. 113–121 (1974).

Skinner, Wickham A. "Manufacturing: Missing Link in Corporate Strategy." *Harvard Business Review, 5* (May-June 1969), p. 20.

CHAPTER 5

WASTE, BY-PRODUCTS, AND SYSTEMS ANALYSIS

The site having been chosen, the need for several new members becomes pressing: persons with local regulatory and environmental expertise and with detailed industrial systems design experience are critical to any effective facility design. New complications created by the new products, especially in relation to unwanted byproducts or wastes, must be considered. A safe environment is paramount for economic and conflict-free operation.

OUTPUTS

Before making the many detailed decisions about how to develop a site and its internal processes, the site development team must take yet another "big picture" look—this time, at what will be taking place when the site becomes active. *Active* in this sense means manned equipment, materials, and other resources entering and products leaving. Doing this analysis effectively is crucial to success in IFM's later, supporting, role because failures at this point mean expensive rework later.

The thrust of CIRM is that of integrative thinking and analysis, and this chapter is no different. The SD team must examine the prospective facility and its processes *systematically,* seeking insight into *all* the "transactions with the environment" that will be taking place under various scenarios. In a sense, the goal of such systems analysis is much like the contingency planning discussed in Chapter 4: trying to make design choices that anticipate operational consequences (especially unwanted ones), so that undesirable effects can be minimized, rather than fixed later.

Although other illustrations will be provided, this chapter emphasizes scrap, waste, and by-products—frequently overlooked aspects of facilities and production planning. These unwanted or unexpected outputs sometimes occur without any associated manufacturing activities. We are all aware that an idling automobile engine consumes fuel, generates heat, emits noise and exhaust gases, and is unlikely to be doing any useful work at the time. Large circulating-water systems don't stop running when marketable output ceases. Plants trying out a new process may generate tons of scrap or rework. And chemical and other process activities can add large quantities of off-grade or unusable products while trying to generate marketable ones during both changeover and routine production.

The reason for this emphasis has been mentioned before, in the context of environmental activism, regulation, and the overall JIT/TQM goal of eliminating waste from the workplace. So-called rustbelt (or smokestack) America provides another motivation. As a number of plants have discovered, there is a potentially enormous life cycle cost if by-product issues are ignored or remedied after the fact. And, bluntly speaking, it is usually the IFM group that gets left holding the bag, and has to deal with such wastes once manufacturing operations begin.

Today's IFM practitioner must learn to deal with all of these outputs proactively and methodically. This approach is called *Systems Analysis*.

SYSTEMS

A system description relates three interdependent elements: the nature, quality, and quantity of the outputs; the way(s) that inputs are altered (transformed) by the process or system to provide the outputs; and the nature, quality, and quantity of the inputs of the system. Finally, the system includes feedback so that changes in any output, transformation, or resource input (expected or actual), can evoke responses, making the system self-regulating.

For example, a manufacturing facility staff member might be interested in the man-machine system at a particular piece of equipment, or perhaps the interaction between material flow and equipment speed. Here, the specified output quantity would probably be the driver for the system, and the equipment, operator time, parts-handling system, and materials would be "sized" against the desired output. The system then could be analyzed for the effects of various changes in any of the parameters (feedback). The study result would aid in equipment choices and in estimating "surge capacity" for handling variations.

FIGURE 5–1
Systems Analysis

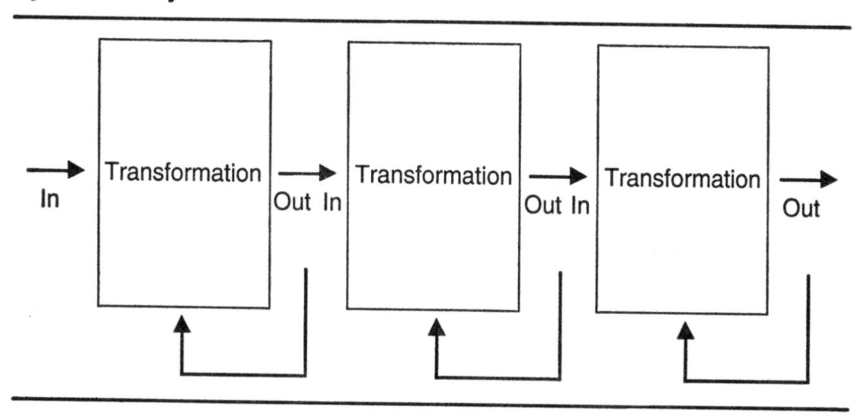

Another kind of system view is involved when the balance of a line of equipment is studied, perhaps for assembly of a product. Desired output rate, number of work stations and people, the kinds of equipment available, and the division of work among the team would be analyzed to seek some sort of lowest-cost solution. In this case, it is not only the performance of the individual machines, conveyors, and people that must be examined, but their interactions as well.

Yet another illustration might be the material flows from a producing facility (supplier) to a consuming facility (perhaps an assembly plant) that require transportation (the process moving between them). Schedules, volumes, distances, and costs would all be involved, with impacts on logistics and facilities planners.

In each illustration, the object is the same: to measure, describe, analyze, and then—using the ability to adjust various factors—to optimize the system in terms of some criterion or goal. This process of systems analysis can be a powerful tool. Detailed studies like these are a part of the basic facilities planning work to be discussed in Chapter 6.

Systems Analysis and IFM

Systems analysis (SA) can be used to examine the interactions among the various internal elements of any system (see Figure 5–1). As part of the facilities planning effort, SA has great value because of its generality. It can apply effectively as part of capacity and facilities network studies or at the single-piece-of-equipment level, as noted. The SD team members have great interest in each area, depending on how far the design process has moved.

The team must examine what manufacturing process steps are involved, especially in terms of rate and flow. This can lead to an examination of capacity at each process step, as well as the planned transfer rate for materials and parts-in-process. The team will also examine distances, materials-handling needs, layout, staff requirements, areas of high energy consumption or generation, the need for and the availability of information, and so on. Quality will also be considered in terms of yield or its complement—fall-out rate—so that the starting batch can be sized right to meet the finished-goods schedule. The various feedback measurements available from these studies include current efficiencies, bottleneck identification, data allowing schedule construction, and probable equipment outages for planned maintenance. Required crew sizes for a specified production volume and mix (or conversely, the practical output for a given crew size) can also be derived directly from a complete analysis. The SD team will be involved in most aspects of this during detailed planning and throughout the many IFM operations support activities.

"BACKWARD" ANALYSIS-OUTPUTS FIRST

Ultimately, any effective system must satisfy its internal and external customers. The facility is the system that must achieve this, leading to output specifications for the marketable products and services demanded. If these system requirements can be clearly understood and translated into the details, then the issues of transformation processes and resource inputs are in the right perspective. Therefore, this presentation began with outputs, rather than the inputs-transformations-outputs format of traditional systems analysis (see Figure 5–2).

Product outputs include primary ones—to be readied for movement to a customer—and secondary ones, sometimes not even considered in analysis. Some secondary outputs are of principal concern to the facilities analyst, including by-products, scrap, heat, human and product wastewater, airborne particles, water-borne contaminants, chemicals, and hazardous fumes.

FIGURE 5–2
An Economic Input-Output Systems Analysis

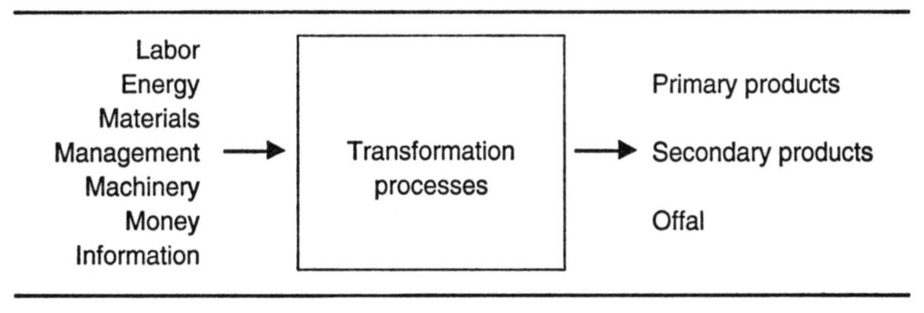

Primary Product Outputs

The charter assignments of products and output rates become the basis for sizing the plant and its production machinery, choosing equipment (based on the manufacturing-process design), determining how to organize the plant (layout) based on the manufacturing concept and the specifics of the various process steps, and examining the logistics of supplier deliveries and finished-product shipments to develop an overall flow.

Secondary Product Outputs

By-products in a woodworking plant, for example, might include the "offal," or short pieces of dried, finished wood that are too small to be converted into *any* finished product desired by current customers. Such pieces are usually destined to become chips or sawdust, to feed the boilers for kiln firings, or to heat the plant needs. But some other shop might be able to use these offal pieces because their product mix includes customers for smaller finished pieces.

An equivalent in a metal stamping plant is the knockout, or smaller piece of strip that is the waste from punching a ring or plate. For example, an electric motor or alternator has a set of plates that are pinned together and magnetized or wound with copper wire to form the field coil. Inside diameters are established to provide clearance for the rotating armature shaft that, interacting with the magnetic field of the outer coil, produces electricity or functions as a motor when fed energy. The inner piece of steel punched out of the strip being fed to the press is waste. But tool redesign can allow a single punch action to create both the field plate *and* the set of plates that are locked onto the armature shaft, to form the armature itself. This cuts down the amount of strip used and the resulting scrap, and the machine capacity is consumed.

There are many other examples of by-product use, ranging from crude-petroleum and natural-gas gathering and refining operations (where the natural gas "strippings" become the basis for the petrochemical industry) to the marketable chemicals derived from "scrubbing" stack gases.

Demanufacturing and Remanufacturing

Some fascinating trends are appearing due to a combination of social consciousness, disposal problems, and economic importance. Audi is designing an automobile than can essentially be run backward through the manufacturing transformation process (*demanufacturing*) to yield component parts, the latter either destined for recycling into raw materials (as Nucor does) or for rebuilding. The hazards of abandoned refrigerators have triggered a demanufacturing emphasis by the Electrolux Corporation. Abandoned buildings, too, can be dismantled for useful components (antique brick or timbers) and the remainder placed into construction sites as fill to raise grade level.

Remanufacturing is also on the rise. Many of us are familiar with engine rebuilding performed when an auto engine throws a rod, for example. Mail order catalog companies like J. C. Whitney carry pages of rebuilt engine elements (e.g., short blocks for many popular engine sizes) from which the customer may order. Automobile alternators, air conditioning compressors, water pumps, brake shoes, and entire front and rear calipers are similarly rebuilt, which is of economic advantage to both the rebuilder and the replacement customer.

Diesel Recon, a subsidiary of Cummins Engine Company, has an entire facility in Memphis dedicated to rebuilding Cummins engines when they are turned in by customers. And Electrolux has for many decades remanufactured their trade-in vacuum cleaners and sold them through regular dealers.

Waste

The naturally occurring residue of transformation processes (e.g., a manufacturing system) gets a backward definition: anything that occurs when making the primary product that does not appear to be currently (or practically) marketable.

Scrap

At ALCOA's Tennessee North Plant, the conversion process from ingot to finished product involves a substantial amount of metal removal, which is typical of process businesses. This occurs in scalping, to take off oxidation prior to hot rolling; during hot rolling, as the thickness of ingots is reduced from almost two feet to one-eighth inch, to remove the "alligatored" ingot ends so they won't allow oxide inclusions in the products as they're later cold rolled; and in cold rolling itself, during end trim and side trim to finished dimensions. The difference between finished-product weight and ingot starting weight is called *recovery.* The trimmings go into the scrap loop, where, carefully segregated by metallurgical properties, the metal scrap is remelted and modified to become another cast ingot. Because of the great volumes, SA effort is always being expended in seeking ways to improve the conversion of scrap into finished product, to improve the initial recovery. Although the largest part of the metal is obviously not lost (though there is furnace vaporization and gaseous loss), there is a repeating consumption of capacity, energy, and labor by the system, which can be saved by recovery improvement (see Figure 5–3).

Heat

Another issue is what to do with the heat required for processes to take place or the heat generated by processes themselves. Steel and aluminum making are obvious examples. Many thousands of gallons of hot water are generated through the quenching processes used to "freeze" molten steel castings as they are readied for the final shaping stages. Casting plants must anneal the rough castings before machining, involving "soaking" at elevated temperatures, then cooling for ease of handling. The large heat expenditure is carried off as hot air through the exhaust system.

Heat Recovery

Electric utilities have large boilers that use fossil fuels for combustion. As the hot gases pass around water-filled tubes in the boilers, the water becomes steam that drives the generators producing electricity. But there is considerable thermal energy left in the hot gases after they have made as many passes over the tubes as is practical in terms of steam generation.

FIGURE 5–3
Systems Analysis of Typical Scrap Loop

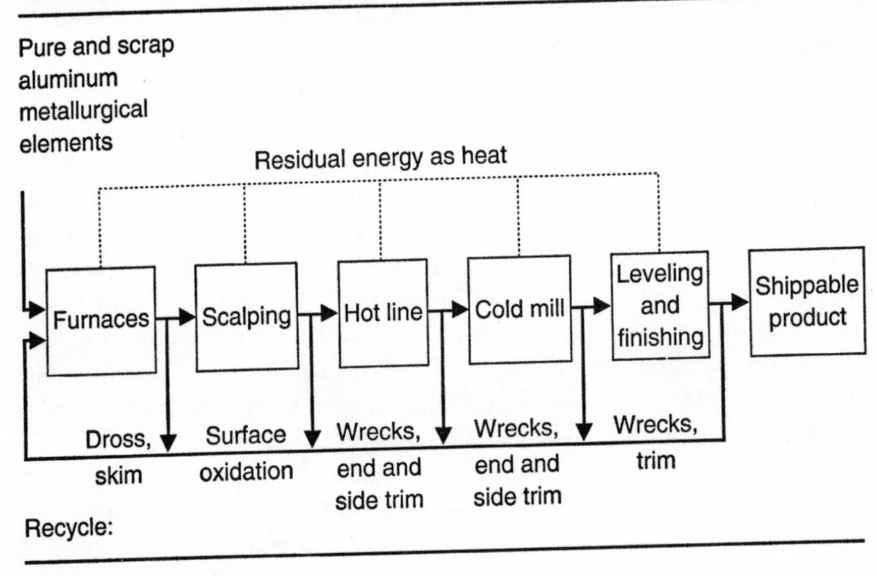

Enter the "economizer" system: tube nests positioned in the exhaust gas stream outside the boiler to heat as much as possible the feed water that will pass into the boilers. Readers might be more familiar with this concept as applied to the automobile turbocharger system and heating-pressurizing of combustion air.

All outputs of a system require comprehensive analysis, especially those relating to energy. The overall cost efficiency *can* be enhanced, along with reducing the incidental costs incurred by having to deal with waste heat (see Figure 5–4).

Water

When water is needed for cooling, there is always a potential for recovery of the heat carried away. The economics of the equipment investment costs compared with the heat saved lead to a measure of how much effort should be expended. This is also true when water must be purchased or when treating wastewater becomes expensive.

When Farmville, North Carolina, officials examined why their water treatment plant needed to expand, they discovered that each of their textile plant customers was using, then redelivering as waste, many

FIGURE 5–4
Systems Analysis of Energy as Waste

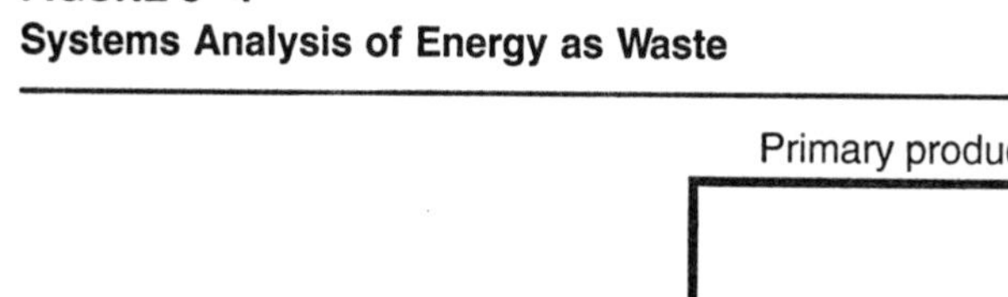
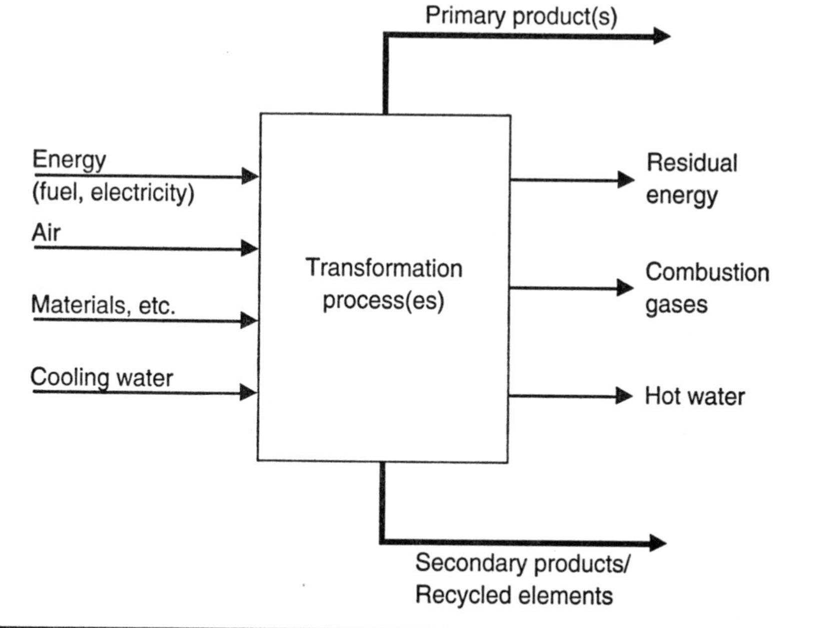

thousands of gallons of water daily. Accordingly, the utility rates were revised to be based on the volumes from each customer to be processed through the water treatment facilities. Alarmed, the more progressive plants undertook process revision. Through cooling and precipitation of the dye wastes, they have been able to recycle up to 95 percent of what was previously wastewater.

Airborne Particles

Most airborne wastes are familiar: asbestos, coal dust, silica, and fiberglass, for example. Some of the less familiar include acid rain, which comes about from the combustion of Bunker C (residual fuel heated, then burned as boiler fuel on shipboard) and other petroleum fuels; paint vapor, arising from spraying operations; and machine shop haze, which arises from the cutting oils and cooling lubricants used with machining operations.

In each case, the challenge is to *anticipate* the presence of the particles, examine the potential health hazard, and deal effectively with the problem. Airborne particles also affect adjacent processes. Clean rooms for the management of some pharmaceutical and semiconductor processes and enclosed and heavily filtered spaces for paint spraying because of particulate contamination are obvious requirements.

Waterborne Pollutants

Human wastes, chemical wastes, and solid wastes in suspension pose significant facilities challenges. Dilution, and in many cases natural cleansing, occurs when industrial wastes can be mixed with large bodies of water. Unfortunately, there are many examples of streams and rivers—even large bodies of water (e.g., parts of the Great Lakes, and the Gulf of Maricaibo) that have become almost uninhabitable except for hardy aquatic scavengers. Only the most stringent regulations and infusions of capital have enabled noticeable improvements in these waters.

Chemicals and Hazardous Fumes

Maintenance of safe standards for the people in direct contact with such processes is only a part of the challenge. There have been a number of reports of dangerously high concentrations of mercury and other metals in fish species sought for food or sport. And fish kills seem to be almost an everyday occurrence. Plating operations have been a continuing source of difficulty because of airborne and liquid chemical waste disposal problems. Designs have moved toward inherently "tougher" metals to bypass plating. These examples reinforce the need for good SA and for much more extensive systems to be analyzed. In environmentally sensitive areas like Florida's Everglades, entire ecosystem models have been developed and new contributions evaluated before issuing permits for roads, buildings, canals, landfill, or industrial-effluent discharge. California and Florida, among other states, demand environmental impact studies prior to allowing residential construction. These studies must include congestion, population, human and industrial waste, and economic-support consequences.

Noise

We've all read about the aural disabilities suffered by "heavy-metal" bands and the teenagers who listen to them. Industrial noise has become the subject of ever closer scrutiny and OSHA regulation, and an increasing number of civil damage suits have been instituted by injured employees.

An interesting solution to the problem of industrial noise has come out of efforts to improve metal-forming operations at JIC/Pikeville. Typically, quadrant gears—formed as part of the operating assembly for raising and lowering car windows— were blanked, pierced, put through final stamping, and then deburred and tumbled to achieve their final shape. Stamping is very noisy, with regulatory requirements for employee ear protection and sometimes enclosures to abate the sound further. A solution arose from the desire to achieve more-precise one-step forming. After making a material change and acquiring new equipment, liquid nitrogen is now used to chill the metal blank. A precision die presses the metal, which, thus embrittled and stressed, cracks to take on the specified shape. The result is an almost noiseless operation, not to mention significantly greater productivity. The formed gear also meets performance specifications without any further finishing or plating!

FIGURE 5–5
Systems Analysis of Wastes and By-products Synthesis

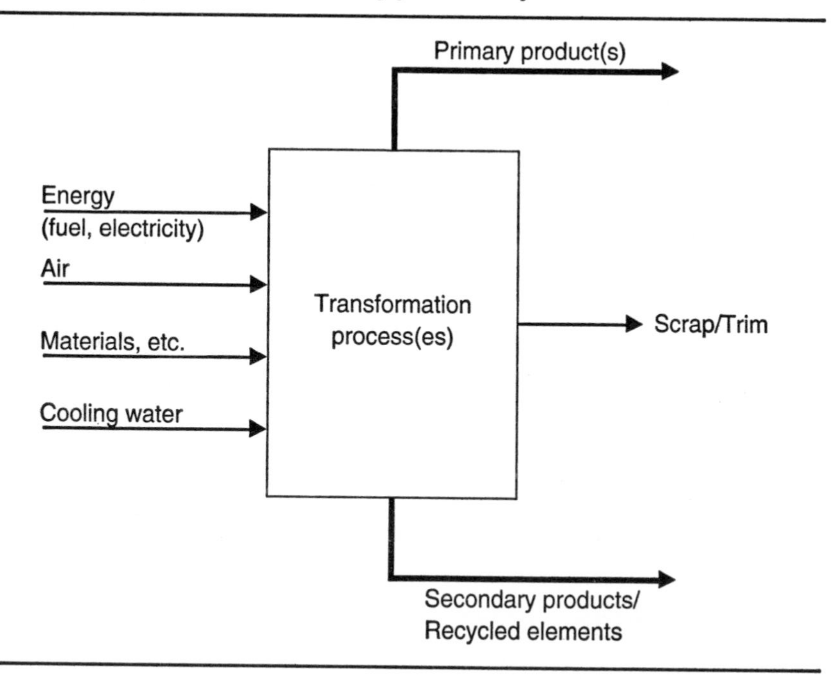

Taken altogether, unrecognized waste and undesired by-products constitute an important challenge to IFM because these problems most often fall to the facilities function for solution (see Figure 5–5). Anticipation and, where possible, solution of these problems through the initial design is much superior to retrofitting equipment with filters, exhaust hoods, and sound deadeners. The difficulties incident to die changes within a noise abatement enclosure have been turning setup and operating personnel gray for years.

By-Product Consumers (Processors)

Glass, paper, plastics, steel, and aluminum are all used as packaging materials for consumer and industrial goods, and management of packaging disposal or recycling is of great concern to the site and to the company. The packaging-material manufacturing processes themselves have significant impacts on the environment, and both compliance and cost management aspects are important. Regulations are becoming more stringent with each passing year, and solid-waste disposal sites are becoming less available. Some states have established "package taxes" on what were once viewed as nonreturnable containers, so that the individual user will be motivated to return the empty for a refund. Unclaimed bottle deposit "taxes" are paid to the state to help with collection costs. This can be a major customer-supplier category for some businesses.

The following are some examples of creative uses of by-products that can convert disposal costs into profits:

- Using furnace cinders and ash for building materials (cinder block)
- Using scrapped automobiles as the feed stock for steel mini-mills
- Melting of used glass and aluminum containers, reforming, and resale
- Stripping out the heavy "ends" of natural gas to become petrochemical feedstock rather than "flaring it off"
- Reprocessing lubricating oils, particularly automobile engine oils
- Segregating various paper grades, gathering, and repulping to reduce new-timber consumption
- Using lumber mill waste for pressed board, particle board, and fuel

Some municipalities have undertaken large-scale waste-processing operations, stripping metals, glass, and some plastics and papers out of domestic solid wastes gathered by the sanitation department. With all currently marketable materials removed, the remainder is burned and the heat used to generate power, which is sold to the local utility. This can have a significant impact on industrial waste classification and processing as well, so IFM needs to have good relationships with the municipal supplier-customer.

According to *IFMA News* (April 1992, p. 5), "Wisconsin has passed what many consider to be one of the most comprehensive and strict recycling legislations in the United States. The 1989 Wisconsin Act 335, which relates to the recycling and management of solid waste, bans a comprehensive range of recyclables from landfills by 1995. The Act established prohibitions on landfilling and burning of white goods, car batteries, and waste oil by 1991, and beginning in 1993, yard waste joins the list. Beginning in 1995, prohibitions on landfilling and burning [a long list] of recyclables. The Act makes landfills liable to fines if they take banned recyclables."

TRANSFORMATIONS

The conversion of inputs through various manufacturing processes to become specific outputs is called *transformation*. This can be done through pressure, heat, cutting, welding together, forming, reduction (as with ore concentration), catalytic reaction (as in petroleum refining), batch blending (as with pharmaceuticals or rubber), or combinations of several processes. Two texts in this CIRM series address transformations in detail, so discussion of this subject here will be brief. Figure 5–6 suggests the variety of processes available.

Process choices reflecting tradition rather than output need create difficulties. The earlier examples of improved motor plate blanking and of JCI/Pikeville's metal-forming improvements show how design specifications can be achieved through a revised set of transformation processes, with gains in other areas.

As another example, joining processes have undergone major change with the introduction of high-strength adhesives. Companies like Textron Aerostructures have incorporated a bond shop (using adhesives for structural joining) in their aircraft component construction, supplanting many thousands of rivets. Bonded composite materials incorporating high strength, light weight, and easy formability are being used for many

FIGURE 5–6
Manufacturing Processes Classification

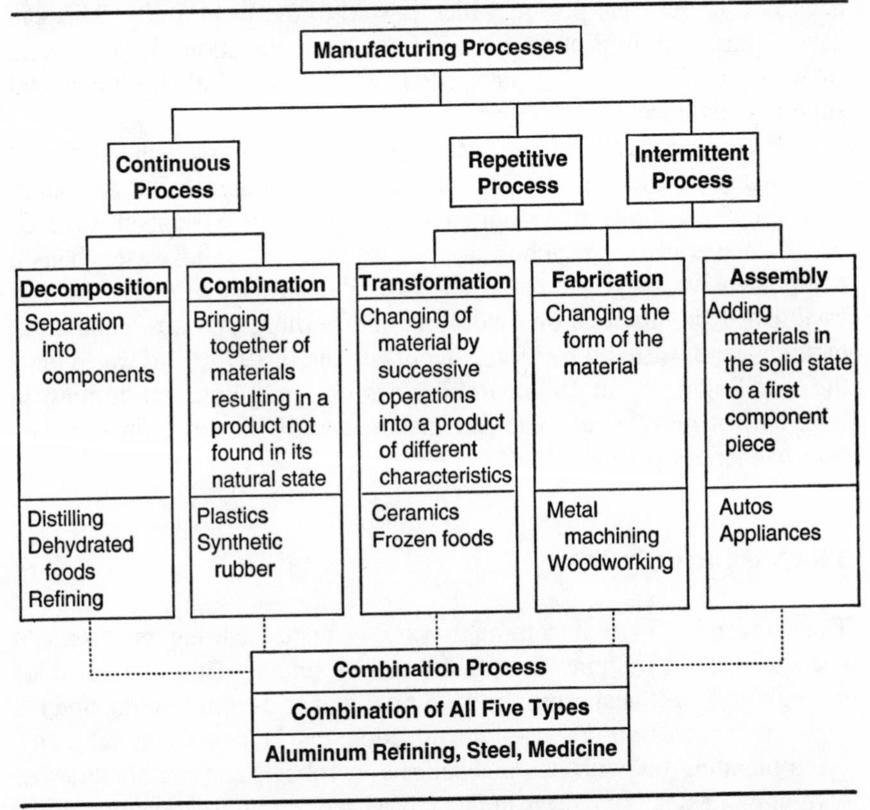

Source: Robert R. Bell and John M. Burnham, *Managing Productivity and Change* (Cincinnati: South-Western Publishing, 1991), Fig. 4–1, p. 70. Reprinted with the permission of the publisher.

structural members that were previously machined from solid high-strength aluminum alone. And one-piece, high-impact plastic moldings are being substituted for joined metal parts, with many weight and cost advantages. The IFM involvement, through SD teams, includes the clean-room atmosphere, personnel protection from vapors and hazards from splashing, extensive exhaust system development, and significant layout modifications to accommodate the new processes and their position in the process flow toward final assembly. Examining the transformation process choices from the standpoint of the needed output function *and* the by-products can yield great dividends.

FIGURE 5–7
Systems Analysis of Inputs for Manufacturing

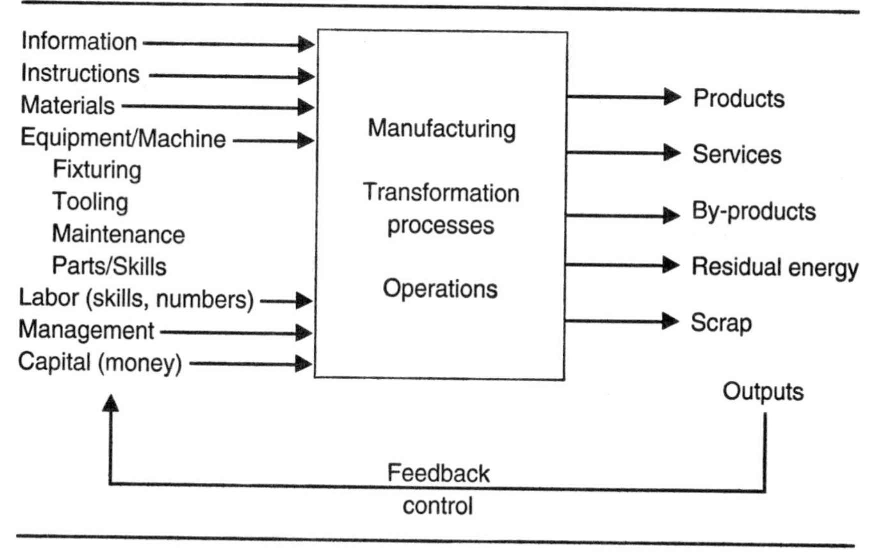

INPUTS

Once a system is defined, inputs are those elements that enter the system by crossing its boundary. For productive systems (e.g., transformation processes), these may be enabling resource factors. Process designers specify equipment and tooling; facilities designers evolve the plant and its physical features; and manufacturing uses what has been provided to carry out the transformation assignments. In sum, inputs will be those resources needed, and those inseparable from resources absolutely needed for the processes to take place (see Figure 5–7).

Instructions

The requirements that the system is asked to perform—for a customer or other parts of the company—become the basis for transformations. This form of input serves to establish timetable, quantity, price, and other factors of compliance or of restrictions on system activities. Instruction in the form of regulation or compliance order can have mandatory elements that may negate some or all plans for production that are otherwise responsive to a customer's request!

Information

Information is a critical resource to a productive system. Knowledge is power. And planning is not possible without sufficient, accurate, and timely information, including, but not limited to, the timing and availability of productive resources, the current capacity and workload, and the future.

Materials

Physical transformation cannot take place without materials. Beyond this, the analysis of a productive system involves materials, both flow rates and total volume. SA considers *all* materials needed in support of production, including repair parts, operating supplies, bulk and unit packaging, lubricants, paints, gas for furnaces, and so forth.

Equipment and Machines

To some people, machinery seems to be part of the transformation process and not an input to it. SA considers equipment an input in that its physical presence in the system does not guarantee its availability to perform the conversion activity. Specific commitment of equipment creates information that allows the scheduling of production activities.

Labor

Skilled operators represent an essential input to productive activities. The work force also accomplishes the conversion of information into activities, the changeover or reorganization of equipment, and the setup of materials to go through the transformation process. People are a literal input and output if the plant site is "the system," because people coming to work and going home cross system boundaries. This physical aspect of people inputs parallels that of materials and applies at the most micro level in the vicinity of a machine, as well as moving onto and off plant property.

Management

Managerial skills also represent an essential input. These skills provide direction to the efforts of others in acquiring inputs, converting information, scheduling the equipment and work force, and communicating with the customer. The more complex the processes or infrastructure, the more managerial time is needed. The skills needed to support teams differ from traditionally directive practices.

Energy

Transformation processes require energy—sometimes, in enormous quantities. Major energy consumers include heating of materials, power for lighting and ventilation, electrical power for running equipment, pumps, fans, conveyors, and lifts. In offices, the same factors are present, although in different volumes and proportions. *All* work both consumes and gives off energy in one form or another, so understanding the energy inputs to a system is crucial to its optimal use.

Capital

This input, like that of management, enables the acquisition of other productive resources and must earn a return on its investment. SA looks at capital in the context of costs, revenues, and profits, applying these measures against each resource and then in their optimal combination, achieving desired outputs through the site development and operating processes.

A Facilities Example

An industrial building might be erected to provide shelter and utilities for a manufacturing operation. It could already exist, having previously served some other manufacturer, or it might have been a distribution center, located close to existing customers. In the aggregate, the analyst team might want to study the entire building in terms of its inputs and outputs and possible materials flow. For a given volume of output of punched parts, for example, there will be an input of a number of coils of steel. If only one part and only one machine is involved, the system could be simple enough. If a number of parts and machines are involved, perhaps they should be considered separately and then all the input requirements added together. The same procedure should be applied to the outputs (see Figure 5–8).

The first task of systems analysis is to enumerate all the significant inputs and outputs. Are there unusual energy needs? Light, heat, and ventilation requirements? Lubricants? Tools or fixtures? What about packaging materials to hold the finished parts? Pallets to move them onto trucks or conveyors to reach the loading point? Are there other outputs beside the packages of parts—for instance, metal trimmings from the punching operation or the banding that held the strip? Or the wooden or steel cores that were used to hold the coil steel? Can noise, vapor, or flying metal chips be a factor?

FIGURE 5–8
Systems Analysis of an Industrial Facility

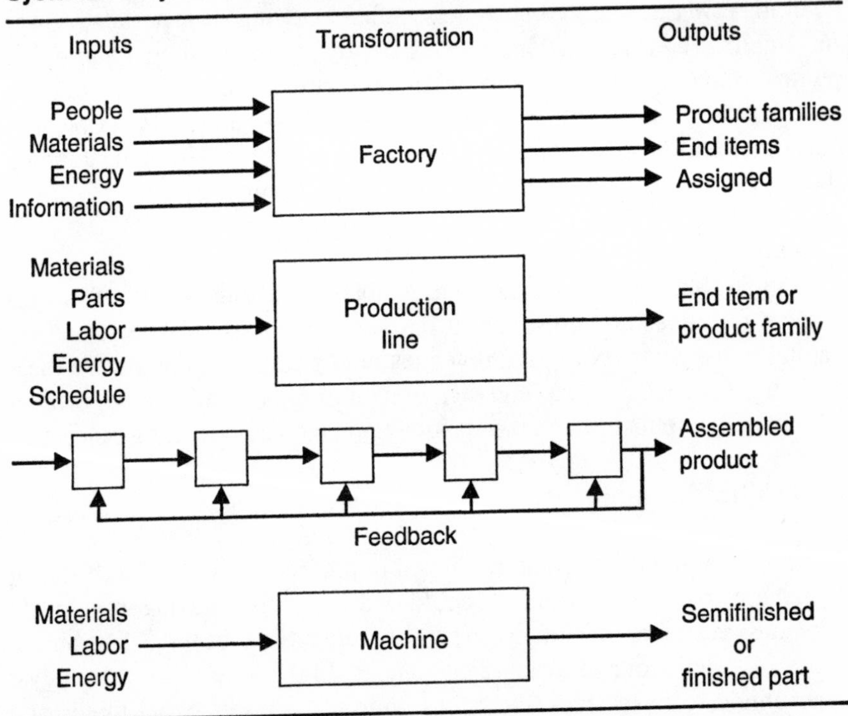

Another line of inquiry about the building might be its suitability for its mission. Will there be a need for large doors or ports to admit materials or vehicles? Will a large amount of natural light be needed? How heavy are individual pieces of equipment, and what kind of supporting structure does each require? Can they be positioned once and for all, like big steel-rolling mills or paper-making machinery? Or will the layout need to be flexible because many different products and processes will be involved over the life of the plant? In the former case, localized foundations may have to be built up; but in the latter, the entire floor area may need to be capable of supporting the greatest machinery weight contemplated. Can production capacity be met through a number of small machines, so that relocations are quick and inexpensive?

In addition to the description of a system, the analyst needs to know its purpose. This can lead to understanding of the criteria by which it will be evaluated—and if there are several, which are the most important.

FEEDBACK

Feedback allows examination of process behavior (performance); depending on the purposes of the system, it is measured at various points. Feedback permits control if the inputs, processes, or outputs may be varied by the observer. Sometimes for machine processes, the feedback is a part of a closed-loop system, and the machine computer controller will automatically adjust the machine to maintain the specified output.

A measurement of actual production units completed would give information about net volume. If compared with the machine load or scheduled manufacturing program, the measurement would show how well the system was meeting the plan of work. If the output were compared with the inputs, some measure of efficiency and of process yields could be determined.

The frequency of output feedback should match the plant manager's concerns about how the plant is doing. Daily briefings could exchange information among department heads, production control personnel, and the product manager, and decisions could be made about adjustments to bring the system into alignment with its targets *or* to adjust the targets to reflect what is possible given the circumstances (machine breakdown, material shortage, etc.). There also may be customer-driven changes in priorities, calling for adjustments to meet the most-current expectations.

All systems can provide feedback. It can occur at various points and levels within the overall enterprise, at different frequencies, and with different degrees of detail, all depending on the needs of the user. Furthermore, data may be transformed to meet different needs. The total number of units produced can be "multiplied" by their bill of materials to reflect material consumption; multiplied by the unit costs for the various resources to show the rate of cost accrual; and by overhead and profit percentages to show projected value of completed units as assets and as marketable products in revenue terms.

Note that in this sense, an *information system* has its own outputs, transformations, and inputs, depending on the needs of its customers. The result of SA is a set of recommendations addressing the proper settings for the elements of the system to achieve its best operations in accordance with the design criteria. This could include layouts, equipment choices, staffing levels, material volumes and points of entry and exit, and inputs and outputs of information for planning, execution, and control purposes.

SYSTEMS CRITERIA

What constitutes a system depends entirely on what the viewer's interest is. What is perceived as internal, and thus reflected in the system description, is contrasted with what is perceived as external, and therefore to be treated as an uncontrollable variable, or perhaps ignored, as is often the case with waste and secondary outputs. Production control uses units of production, while the plant manager may also need costs. Division management wants costs, but may also need revenues for units sold. Corporate management may need both revenues and costs, and will want profits and profitability as well. A general system for manufacturing follows as Figure 5–9.

Another important viewpoint is that of purposefulness. Is the system free to take on any form that will meet its output requirement (that is, is it customer driven)? Or is the objective to get the greatest possible output from limited equipment (transformation ability) or from the limited resources available as inputs to the system? A system designed for efficiency may disappoint a customer. Conversely, a system designed to please a customer may not be able to operate at maximum efficiency. Choices to meet one criterion are unlikely to meet others optimally.

Finally, the systems description and analysis can be used to optimize the current situation or to define changes needed to meet changed conditions. *Optimum* for the current situation might be in terms of productivity, leading to *removal* of some of the inputs (materials, people, or materials) while meeting output targets. For other conditions, the optimum might entail the minimum *additional* resources to meet changed volumes, mix, or quality requirements. In either case, the SA process is now in a normative—as contrasted with descriptive—mode and seeks to achieve improvement.

Needs for Analysis

IFM is often left holding the bag because of insufficient information or time for study and contribution to the most effective solutions, so this chapter has focused primarily on waste and by-products analysis. There are many other important areas where this output-transformation-input methodology can be helpful.

Site Selection. Site selection includes customers, suppliers, other plants, costs, financing analyses, and logistics.

FIGURE 5–9
Systems Analysis of the Overall Synthesis for Manufacturing

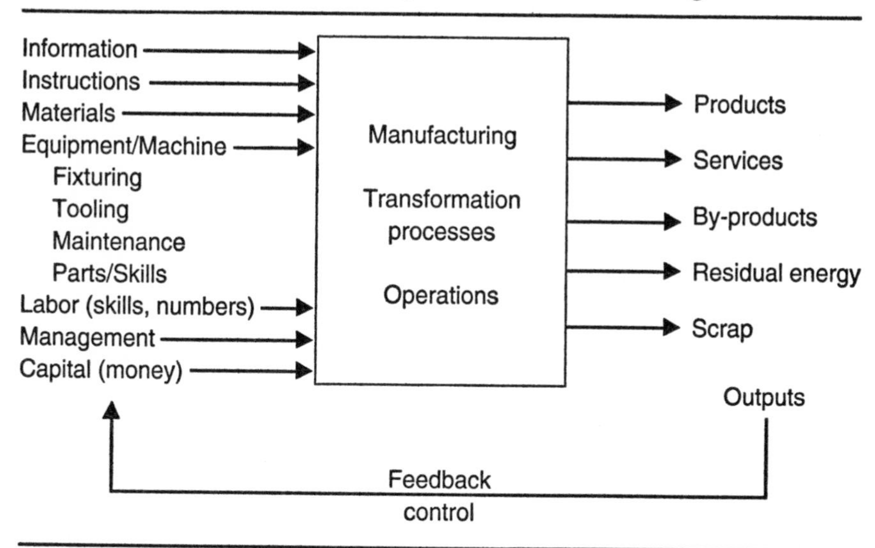

Transport, People, and Materials Flow: Macro. Includes the overall plants system, transport linkages, and the detailed logistics system. At the specific site, there's a need for building-orientation analysis, considering human access and egress, truck routes, and materials and utilities entry.

Equipment Layout, Materials Handling, and People: Micro. These systems must be integrated, but they must be analyzed separately for some issues.

Energy Management. SA is required for not only the manufacturing products and processes but the general facility heating, cooling, and ventilation.

Structure, Organization, Equipment, and Materials Handling. With tentative selections and decisions made, SA addresses the specifics of flows, support, and costs.

Mix and Product Changes. SA can assist with both known and probable changes to production assignments and examine the best transitions from existing plant to any new configuration.

Maintenance Strategy and Tactics. With a given layout, equipment inventory, and manufacturing concept, SA can examine both the overall strategy for optimal equipment availability and the means for supporting the capability of individual pieces of equipment.

MRO Inventory Management. Locations, quantities, and means of records maintenance can be planned using the APICS-developed systems, with the current supplier restocking programs factored into the planning system.

By-Products and Waste Management. Once the presence and quantities of these outputs have been defined, effective disposal means can be planned.

Cost Projections and Related Revenues. Financial SA can be used to evaluate both stand-alone (product life cycle) and plant life cycle equipment investments, relating them to product revenue projections.

LESSONS FOR THE INTEGRATED RESOURCES MANAGER

This chapter has focused on methods for describing, analyzing, and improving production systems, with particular emphasis on waste and secondary products. IFM professionals can contribute analysis in strategic and tactical planning through their role in site development teams.

A new rubric has been suggested for approaching SA: start with the necessary *outputs* of the system when dealing with a design or modification issue. That is, simple systems can be created to examine the ties between a fabrication "supplier" and an assembly "customer" just as well as those between the plant as a product supplier and the industrial customer or retail distribution system. If (and only if!) the details of the system output are understood, the development of the transformation processes and the specification of inputs can be effectively accomplished.

"If you don't know where you're going, any road will get you there!" This is all too frequently the case with even simple systems problems. What seems to happen is that *procedures* are substituted for the more difficult task of *analysis,* with the result that actions are based on symptoms, rather than on the underlying problems. Examples include capital equipment justification on a stand-alone basis, facilities "cost efficiency'' based on cube utilization, and ''bottleneck analysis'' that ignores other parts of the process system. The apparent inattention to waste and by-product reduction and disposal is another unfortunate example.

The key to effective systems analysis is that of learning to manage variety. This means that the IFM analyst must become selective and provide focus on what's important *for the purposeful system,* not just an exhaustive treatment of all conceivable factors that have some direct or indirect interaction with *some* aspect of a related element or system. And because there may be several purposes, a number of different interested parties, and many criteria, it may be necessary to do several analyses of the same physical system and to transfer bits of information back and forth among the several ''focused'' systems to make sure that all *relevant* details are present to allow system optimization.

It's important for the IFM leader or site development team member to keep in mind that thorough, creative SA in the examination of process alternatives can lead to worthwhile changes for overall facility success. The SD team must include product and process engineering, manufacturing, and operations personnel, along with customer and supplier representatives, if it is to be effective.

APPENDIX: IFM SYSTEM PLANNING TOOLS

A brief description of a variety of useful IFM/SA techniques, inputs, and outputs is given below. For a detailed discussion of the methods, see the reference list on page 135.

Location Factor Scoring. Proposed plant sites can be ''scored'' by developing fractional weights that correspond with the priorities of the selection team and then scoring each location on each factor. When the scores are combined with the weights, a tentative selection can be made and further studies undertaken on logistics, markets, and so forth. [Schmenner]

Production Energy. Very high energy requirements may limit production locations because of the need for reasonable energy costs and availability of competing suppliers or energy classes. [Berrie]

Heat, Power Sources, and Consumers. Economies can be combined with use volumes and alternative sources to yield operating-cost inputs to public-utility equipment selection. Airborne particles or heat pollution may also be factors. Mathematical programming models are often used to simulate both daily load and overall demand matching. [Berrie]

Transportation Systems and Models. For a specific set of supply locations, product costs and availabilities, and transport costs, mathematical programming can be used to develop the lowest-cost distribution patterns for any given set of markets and demands. [Daughety; Hall]

Cost-Volume-Profit and State-of-the-Market. Linear cost-volume-profit models can develop inputs to equipment selections and scale-of-operations studies. [Altman; Black]

Process Flow Charting. Used throughout this chapter, flow charting—one of the routine steps in SA—describes a system in terms of relevant inputs, transformations, and outputs. Once described, the system elements can be quantified, analyzed, and optimized. [Daughety; Hare; Johnson; Martin]

Manufacturing Concept. Choices of the planning and control procedures for managing production lead to specific and general decisions about plant organization, equipment, skills, and supervisory techniques to achieve design performance. [Bell; Hayes & Wheelwright]

Capacity Evaluation. The usual aggregate studies employ CVP financial analysis and statistical assessments of industry demand to determine the needed capacity. Detailed studies examine specific product fabrication and assembly processes using flow charting, equipment performance data, and output requirements to assess the adequacy of capacity and how best to adjust it. [Gessner; Hayes, Wheelwright, & Clark; Schonberger]

Scheduling Alternatives. The desired timing and volume of product outputs from a system in a specific time frame become the goal of the production schedulers and of the manufacturing system. Mathematical programming and spread sheet simulation packages are in current use for schedule development. [Gessner; Hall]

Line Balancing. At a more detailed level, SA can examine the tasks and times required to assemble a product and subdivide them for optimal load across the entire line, while achieving the desired output rate. Computer modeling and solutions are frequently used. [Gessner; Hayes & Wheelwright; Schonberger]

Ergonomics and Man-Machine Systems. Time-and-motion studies, flow charting the order of tasks, then examining the sequence and effort of the associated human motions can develop effective workplace design. The equivalent for combinations of equipment and operators working together can aid in capacity, scheduling, and balancing studies and avoiding human strain. [Niebel, 1988; Zandin]

Environmental Control. Energy management systems for heating, ventilation, and cooling are used to assess sources, quantities, and the timing of demand. They seek to optimize numbers, sizing, and locations of environmental control systems to meet design objectives. [Greenwald]

Testing and Evaluation Systems. System outputs support specific goals for equipment availability and capability, maintenance for operator safety, specific or general product quality, and regulatory (OSHA, EPA) compliance. SA is used for analysis, examination of the effects of in-line test equipment, and development of computerized databases and retrieval models for planning and for compliance reporting. Cause-and-effect diagrams are another important form of systems analysis, used for problem solving in quality and in maintenance diagnosis (fault-tree). [Hartmann; Hayes, Wheelwright & Clark; Herbaty; Niebel, 1983]

Machine Failure Prevention. System "output" is the required equipment capability, shown through measurements. System "transformation" is the response of the equipment and work piece during operations. Study of critical parts of the equipment/process leads to equipment modifications or changes in tooling on a preventive basis before performance degrades. [Hayes, Wheelwright & Clark; Herbaty; Niebel, 1993; Schonberger]

REFERENCES

Altman, Edward I. *Financial Handbook*. 5th ed. New York: John Wiley & Sons, 1981.

Bell, Robert R.; and John M. Burnham. *Managing Productivity and Change*. Cincinnati: South-Western Publishing, 1991.

Berrie, T. W. *Power System Economics*. London: Peter Preagrinus, 1983.

Black, James H.; and Frederic C. Jelen. *Cost and Optimization Engineering*. New York: McGraw-Hill, 1983.

Burnham, John M., and Ramachandran Natarajan. *Manufacturing Processes Study Course*. APICS CIRM Program. Falls Church, Va.: APICS, 1992.

Daughety, Andrew F. *Analytical Studies in Transport Economics*. Cambridge, England: Cambridge University Press, 1988.

Gessner, Robert A. *Repetitive Manufacturing Production Planning*. New York: John Wiley & Sons, 1988.

Greenwald, Martin L. *Residential Energy Systems and Climate Control Technology: Operation and Maintenance*. Englewood Cliffs, N.J.: Prentice Hall, 1988.

Gross, Charles A. *Power System Analysis*. 2nd ed. New York: John Wiley & Sons, 1986.

Hall, Owen P., Jr. *Computer Models for Operations Management*. Reading, Mass.: Addison-Wesley, 1989.

Hare, Van Court, Jr. *Systems Analysis: A Diagnostic Approach*. New York: Harcourt, Brace, and World, 1967.

Hartmann, Edward. *Maintenance Management*. Norcross, Ga.: Institute of Industrial Engineers, 1987.

Hayes, William H.; and Steven C. Wheelwright. *Restoring Our Competitive Edge: Competing through Manufacturing*. New York: John Wiley & Sons, 1984.

Hayes, William H.; Steven C. Wheelwright; and Kim B. Clark. *Dynamic Manufacturing: Creating the Learning Organization*. New York: Free Press, 1988.

Herbaty, Frank. *Cost-Effective Maintenance Management: Productivity Improvement and Downtime Reduction*. Park Ridge, N.J.: Noyes Publications, 1983.

Johnson, Richard A.; Fremont E. Kast; and James E. Rosenzweig. *The Theory and Management of Systems*. 2nd ed. New York: McGraw-Hill, 1967.

Martin, James; and Carma McClure. *Diagramming Techniques for Analysts and Programmers*. Englewood Cliffs, N.J.: Prentice-Hall, 1985.

Niebel, Benjamin W. *Engineering Maintenance Management*. New York: Marcel Dekker, 1983.

Niebel, Benjamin W. *Motion and Time Study*. 9th ed. Homewood, Ill.: Richard D. Irwin, 1993.

Schmenner, Roger W. *Making Business Location Decisions*. Englewood Cliffs, N.J.: Prentice-Hall, 1982.

Schonberger, Richard J.; and Edward M. Knod, Jr. *Operations Management: Improving Customer Service*. 4th ed. Homewood, Ill.: Richard D. Irwin, 1991.

Zandin, Kjell Bo., *MOST Work Measurement Systems*. 2nd ed. New York: Marcel Dekker, 1990.

CHAPTER 6

DETAILED FACILITIES PLANNING

The product and process designers, industrial and manufacturing engineers, operations and maintenance planners, human resources experts, cost and schedule estimators, and regulatory experts all must work together in the development of the detailed plans and specifications for the site development and the production and warehouse facility contemplated. Everything is still on paper, but this is about to change.

FACILITY PLANNING GOALS

> Facilities planning for the production function is extremely important because of the economic dependence of the firm on manufacturing performance. Since production is value-adding, it receives considerable attention from upper management . . . if problems develop. Unfortunately, production facilities requirements seldom receive adequate *planning attention.**

This chapter focuses on the creation of a detailed (biddable, executable) facility plan for a specific mission at a specific location. Achieving lowest life cycle cost entails broad-based contingency planning (Chapter 4) with the right blend of fixed capital and product-specific capabilities, both to fulfill the initial mission and to extend over the economic life of the site.

An initial facility plan will seek the optimum for all aspects of the manufacturing system it supports. But when the external and uncontrollable

*James A. Tompkins and John A. White, *Facilities Planning* (New York: John Wiley & Sons, 1984), p. 383.

factors are considered, an inflexible design ultimately will be unsatisfactory. Both external and internal changes can require adjustments: JIT, TQM, TPM, and plain old continuous improvement will lead to many changes. The combination of structural provisions, modularity and utilities flexibility, suitable equipment choices, and a strong site team will make the transitions quick, effective, and relatively low cost.

SITE DESIGN

Once the site has been chosen through application of the strategic factors discussed in Chapters 3 and 4, detailed site design can begin. The design approach is most easily seen in greenfield site development, but it can be profitably applied to grayfield situations too. In both cases, the SD team will focus on conversion of the analytic results of using big-picture assessment tools of Chapters 4 and 5 into the detailed bidding and construction plans and specifications.

Once building orientation on the site is settled, the SD team must determine the number of receiving and shipping docks based on plant volume, transport mode, number of different suppliers, transport companies, and distribution vehicles and whether arrivals are scheduled or random. The configuration of delivery and pickup vehicles will also affect all receiving and shipping facilities.

What Constitutes a Good Facility Plan?

Taken quite literally, a good plan reflects the thoughtful evaluation of both known and suspected factors affecting the facility in the initial phases of site development and will be executable on time and within budget. The rest of this chapter addresses those known and suspected factors and ties them into plan development.

Overall facilities planning can be functionally subdivided into the major areas of location and design. Proximity to markets or to vital resources (materials, energy, or, increasingly, skilled people) can make a considerable difference to economic viability. Bricks-and-mortar costs are determined by materials and labor costs in the area plus any special provisions needed. Within design are the structure, services, layout, and materials-handling systems. The unique interactions among equipment and processes, layout and manufacturing concept, and the associated materials-handling system, call for careful site development.

Site-Specific External Factors

There are both external and internal considerations to the plan and to the success of the plant. The uncontrollable external elements and factors will be discussed here in the context of the enormously increased cost of late design changes.

Regulation

Two federal agencies—Occupational Safety and Health Administration (OSHA) and the Environmental Protection Agency (EPA)—operate to protect both domestic employees and the environment.

OSHA was established by statute "to assure as far as possible every working man and woman in the nation safe and healthful working conditions and to preserve our human resources."

In compliance with OSHA standards, firms must provide a place of employment free from recognized hazards involving heat, sound, electrical shock, mechanical devices, dangerous sparking, paint fumes or other noxious gases, hazardous materials, and harmful processes. There are regular and unannounced OSHA inspections. Accidents, complaints, or changes in the standards can all result in OSHA requirements that the employer take action to upgrade protection for employees and sometimes to pay the consequences in terms of fines or even shutdowns.

The EPA was established to protect the environment. Actions by the EPA can lead to restrictions on operating power levels for nuclear or conventionally fired electrical generation. The EPA can restrict the composition and volume of industrial stack gases; sets a nationwide limitation on new autombile exhaust emissions; and provides wastewater purification standards and hazardous-waste regulation.

State and local environmental regulators also have jurisdiction and may, where the situation warrants, exceed federal standards. California exhaust emission standards are more stringent than those of many other states, and both oil exploration and production are carefully controlled. Most states require periodic inspections for all vehicles, and many include exhaust gas monitoring. Florida has imposed stack emission maximums on manufacturers and on public utilities, leading to the use of considerably more natural gas for boilers and steam generators than elsewhere in the United States.

Responses

While IFM has both the capability and the responsibility to *react* to citations from OSHA and EPA, the facility *planning* activities can do a great deal to reduce or eliminate the causes bringing such citations. Thorough systems analysis of all the planned inputs and outputs of the facility and its processes is essential for both community relations and LLCC.

Incentives

Industrial facilities are desirable citizens for a state or locality. Employment, a broader tax base, building and road construction contracts, and employment taxes are among the reasons that economic and industrial development commissions are present in many states and townships. These groups seek to demonstrate to the company the desirability of building a facility in the area and will offer a variety of training and even tax incentives to the new corporate citizen—clearly a positive factor in both site selection and the estimation of costs and potential profits.

Energy

In terms of both availability and cost, energy is an external, uncontrollable factor. The Aluminum Company of America (ALCOA) offers a good example of how important and beneficial energy planning can be.

ALCOA's North Plant (fabricating coil stock to customer specification) is typical of many process plants, with energy representing roughly 10 percent of the cost of casting and rolling. Recently, ALCOA has modernized the plant to address the issues of internal and external scrap (used beverage containers, or UBC). This system was illustrated in Chapter 5 (Figure 5–3). A large investment in statistical process control (SPC) and a move toward total quality has led to rework reduction, improving both the usable capacity and energy consumption. With UBC, the energy required for cleaning and remelting is considerably less than that required for the electrolytic process that creates the pure aluminum from enriched feed stock. Other efforts focus on the energy required for heating for the next process step(s). In some cases, ways were sought to eliminate reheating altogether through creative metallurgy.

Community

From both the regulatory and good-relations standpoints, the entire facility must respond to the local social and political environment. Recall the way Hewlett-Packard went through the site selection decision, resulting in homogeneous, attractive designs that can be found all around the globe. As well as being the place where employees must spend 40 or more hours each week, the facility is the company's face to the world. As a large consumer of various services (sewers, roads, power, water) and thus subject to local revenue-based decision making (taxes and fees), the industrial facility is also a target for expression of any community unrest.

Employees and Families

Being on the payroll does not cause families to stop being local citizens, with all of the opportunities for both good and bad that this entails. The overall design of the facility, its lighting, heating and cooling, ventilation, health services, child care services, bathrooms, storage for personal belongings, equipment arrangement, concern for safety, waste treatment, and desire for a quality product are all expressions of how the company feels about the production team. Similarly, the grounds and the physical face of the plant influence both employees and the local community. A plant's good reputation can attract other qualified people to the area to seek work there.

INTERNAL FACILITIES PLANNING DETAILS

Whether an SD team exists or not, significant interactions should be taking place among the many functional specialties within the company. The following are some of the areas for seeking common understanding and agreements.

Marketing

In addition to the market presence issues affecting location, IFM must know the nature and size of the product, the mix, and the customers' preferred shipping units. These considerations will all affect equipment choices, layout, and the materials-handling design. Competitive requirements like customer service levels, field-based spares, and delivery lead times can also affect facilities details and the production planning and control infrastructure.

Logistics

Global materials movement requires globalization of the logistics function. When the SD team undertakes facility planning, both internal and external simulations may be needed to examine just how materials must flow through plant receiving and shipping areas, warehouses, and distribution centers. The same attention must be paid to the supplier network for maintenance, repair, and operating (MRO) stores, as well as production materials, and for multiple (e.g., Oster's multiplant network) and single sites. Some corollary logistics issues arise from the decision to go to a new greenfield location, to carry out expansion at a current site, or to acquire an existing facility from another division or company.

Manufacturing and Manufacturing Engineering

The degree of integration will affect equipment, tooling, layout, and communication requirements. The nature of the equipment, the degree of automation, and its conceptual organization will affect layout, maintenance planning, materials handling, and both sizing and detailed utilities support.

Equipment

Equipment is functionally determined through manufacturing processes design, selecting the most appropriate equipment for capacity, speed, and precision, as well as the range of parts/work pieces to be handled. Manufacturing cells usually seek dedicated equipment so that each standard machine can be fixtured and tooled to match the needs of the part family or product. Product-oriented fabrication and assembly lines—or production units, mini-plants, focus facilities, or plants within a plant (PWP)—are dedicated to a product family, and the production line becomes balanced to the mix and volume for that product group.

Production and Inventory Management (PIM)

The implications of the manufacturing concept and how production work will be managed affects the size and location of storerooms, work-in-process (WIP) areas, and finished-goods warehousing and shipping areas. PIM, jointly with marketing, must address how to manage aggregate planning (chase or level strategy, seasonality and variability of demand) and typical WIP planning issues of buffer stocks, lot sizing, the production and transfer batches associated with synchronous manufacturing, and so forth. The SD team must have a good vision of how the plant will be running in order to effectively carry out the detailed IFM planning activities. And if—as may be true for most plants in the future—there will be a move toward continuous improvement and JIT implementation, then their ongoing role requires the SD team to make the proper design choices for current and future cost-effective operations.

Money and People

The concerns of marketing, manufacturing, and production control affect the facilities details directly and indirectly, through the effects on the financial and human resources needed to begin production. The typical facility feasibility study carries out break-even analysis based on the sum of fixed and working-capital elements and the variable costs of production. Therefore, inventory investment will not only affect working capital but also the plant layout and space requirements for raw or WIP storage. Similarly, the issues of fixed and product capital affect the ROI calculations and the equipment choices themselves. How the plant will run affects costs and work space layouts, and both the micro-level machine-person interfaces and the macro-level staffing requirements—the latter affecting estimated operating costs as well as spaces for the human resources and training functions.

Technology

In addition to contingency considerations, the SD team needs to incorporate in the detailed plan any known future changes to supplier, production, or customer technologies, particularly in the ways that these may affect internal materials flows or volumes.

For example, the multitude of changes that took place while Xerox addressed regaining its position as a world class company had enormous impact on all portions of the manufacturing system. Product design-to-cost goals (or time, or assembly, or manufacturing cycle) led to new support design goals for the facility.

Summary

Integrative product design, with manufacturing and production engineering involvement, has had great impact on both new-product introduction lead time (time to market) and total introduction cost. A similar opportunity exists to improve manufacturing performance through careful, integrative facilities planning in the design stage. This does not ignore the benefits from continuous improvement of the operating plant nor the mandate that IFM and the SD team support these changes. But just as most of the product costs are determined very early in the design process, so also are those related to the plant itself. And while engineering changes can be accommodated in modifying the product or the facility *after* manufacturing has begun, this is clearly not the preferred approach. Other texts in this CIRM series address these internal requirements in greater detail.

THE FACILITY PLAN

Now the SD team can start to work on the details of the approved preliminary plan. The major factors to be considered are shown in Table 6–1.

Location means more than the plant's role in the logistics and manufacturing system. IFM has two roles: representing and enabling, as the plans evolve. Integral with the plant mission/charter, location includes orientation on the site (once the site has been chosen to fit the system plan) and how the supplier and distribution traffic will best move to assigned drop points. It also includes design of parking areas, lighting, outside facility amenities, recreation areas, handicapped access features, health services, child care spaces, and how food services will be supported. And because the plant's life may well be more than 20 years,

TABLE 6–1
Details of a Facilities Plan

External	*Facility plan*	*Internal*
Regulation	Location	Logistics
OSHA	Structure	Manufacturing process development
EPA	Layout	Manufacturing
Energy	Material handling	Quality criteria
Insurance	Operations planning	Safety
Community	Maintenance planning	Plant environment
Employees/Families	Security planning	Product characteristics
Customers/Competition		Manufacturing concept
Suppliers		

Adapted from John M. Burnham and Ramachandran "Nat" Natarajan, *Manufacturing Processes,* Student Guide (Falls Church, Va.: APICS, 1992), Figures 7–9, 7–10. Reprinted with the permission of the publisher.

location aspects must also consider future needs for expansion, realignment of tasks, and downsizing. Thus, location in terms of strategic site selection and of positioning on the site is extremely important.

A time-based approach to site and structural design might be: Who shall join the SD team? When should the team begin work? What support will SD need? How long will it take to close in the buildings so that equipment can be received? When is start-up expected? What environmental planning guidelines apply?

Project management techniques for providing a systematic way to plan and control the process of site development will be detailed in Chapter 7.

Facility Design

Facility design is a team effort, addressing three elements: structure, layout, and materials handling. Depending on how the facilities plan was developed (centralized or decentralized) and the amount of capital involved, there may be significant representation on the team from top corporate staff.

Structurally, the building must meet the physical conditions surrounding its intended use. Building floor loads and clear ceiling heights will depend on the nature of the business and the equipment to be used. The production lot size of the various items, how these will be moved within the plant and so on, will have an effect on the structure. To allow point-of-use delivery, curtain walls or roll-up sections may need to be provided throughout the building. Design should also anticipate changes in volumes, product mix, and equipment layout and both expansion and contraction. For one AT&T facility, "logical" penetrations for smooth material flow required corporate review because of the potential adverse effect on appearance.

The key to successful structural design is an understanding of how the facility must *function* in terms of products, processes, and mission. Design can be guided by study of other, similar facilities. However, history may be a poor guide if the manufacturing concept has changed from general-purpose to focus, from traditional-flow to JIT (e.g., AT&T), or from make-to-stock to make-to-order but with short lead times.

Layout and Materials Handling

Because JIT material movement is in small quantities—with very little raw or WIP inventory or finished goods present in the system—*focused facilities* operating under JIT discipline are more compact than traditional ones. In addition, focus facilities have closer machine spacing, and less aisleway because of smaller move quantities and more homogeneous parts and products. This tends to negate "sizing" models based on traditional manufacturing systems.

Feedback is quick and make quantities are small, so operators can correct problems immediately. Quality is better and less space is needed for rework. Open, well-lighted facilities will feel uncrowded and still have a very large percentage of total floor space devoted to manufacturing. Because of the many benefits of smaller facilities, the average plant size has decreased by more than 50 percent since 1970.

Activities

IFM must detail the various activities in the overall facility—in offices, receiving and shipping areas, bathrooms, and storerooms, as well as in manufacturing—to make a rough cut at overall plant size based on the area required for each. Space, lighting, and ventilation standards exist for most classifications of activity and can be used for preliminary plans.

Logic of Activities and Space Requirements

As the industrial facility begins to take shape, three interdependent aspects of physical organization of the workplace are involved: the *space* that is required for various activities; the *relative organization* of various spaces to each other; and the details of *layout* that establish the overall plant flows, materials movements, and effectiveness in support of the mission.

With an overall materials flow and processing schematic in mind, the possibilities for materials-handling equipment, lighting, power, drains, and equipment utilities and for other such *systems* as compressed air, lubricating or cutting oils, and scrap (trim, flash, cuttings, grinding debris, airborne paint) can be visualized and standards applied to rough-size the areas. Space required for operator changeover, for maintenance, and for managing the production processes follow from concept and layout. Considerations for space requirements are listed as Table 6–2.

TABLE 6–2
Space Requirements for Equipment, Materials, and People

Equipment Footprint Access Maintenance	Maintenance Tools Lubricants Parts
Machine travel	Materials areas/Handling system
Robotics (arcs) Swing distances	Support services
Operations and changeovers	

Source: James H. Tompkins and John A. White, *Facilities Planning* (New York: John Wiley & Sons, 1984). Reprinted with the permission of the publisher.

Aggregate Materials Flow

For a given set of processes, volumes, and related equipment, individual workstations and the determination of gross space needs offer a starting point. Industrial engineers have been trained to do this department by department, and then, using computer-driven programs, to work up the space assignments to optimize overall layout based on the transactions between them. This approach applies to textiles plants and to steel or aluminum mills like Nucor or ALCOA, where major foundations, materials-handling equipment, power supplies, and utilities *must* be optimally developed for the life of the plant. Some diagrams of how to approach this process are shown in Figure 6–1.

Textron Aerostructures uses a combination of all the layout alternatives shown in Figure 6–1 in its work as a major airframe subcontractor (tail assemblies and wing sections) to the military and to Gulfstream, Airbus Industry, and other commercial customers. The large structures begin in heavy machining (process departments); then move to small subassembly (product family departments) and large subassembly, such as the bond shop for adhesive joining or riveting assembly (production line); and then to final assembly, where all of the parts converge (fixed location) and the end item is created, inspected, and shipped.

The fixed-location department is typically associated with a large, awkward structure where it is more economical to bring the materials, equipment, and work force to the location than to move the product in assembly line fashion. Fixed-location departments at Textron include an Airbus wing assembly or the construction of an office building.

FIGURE 6–1
Plant Layout Alternatives

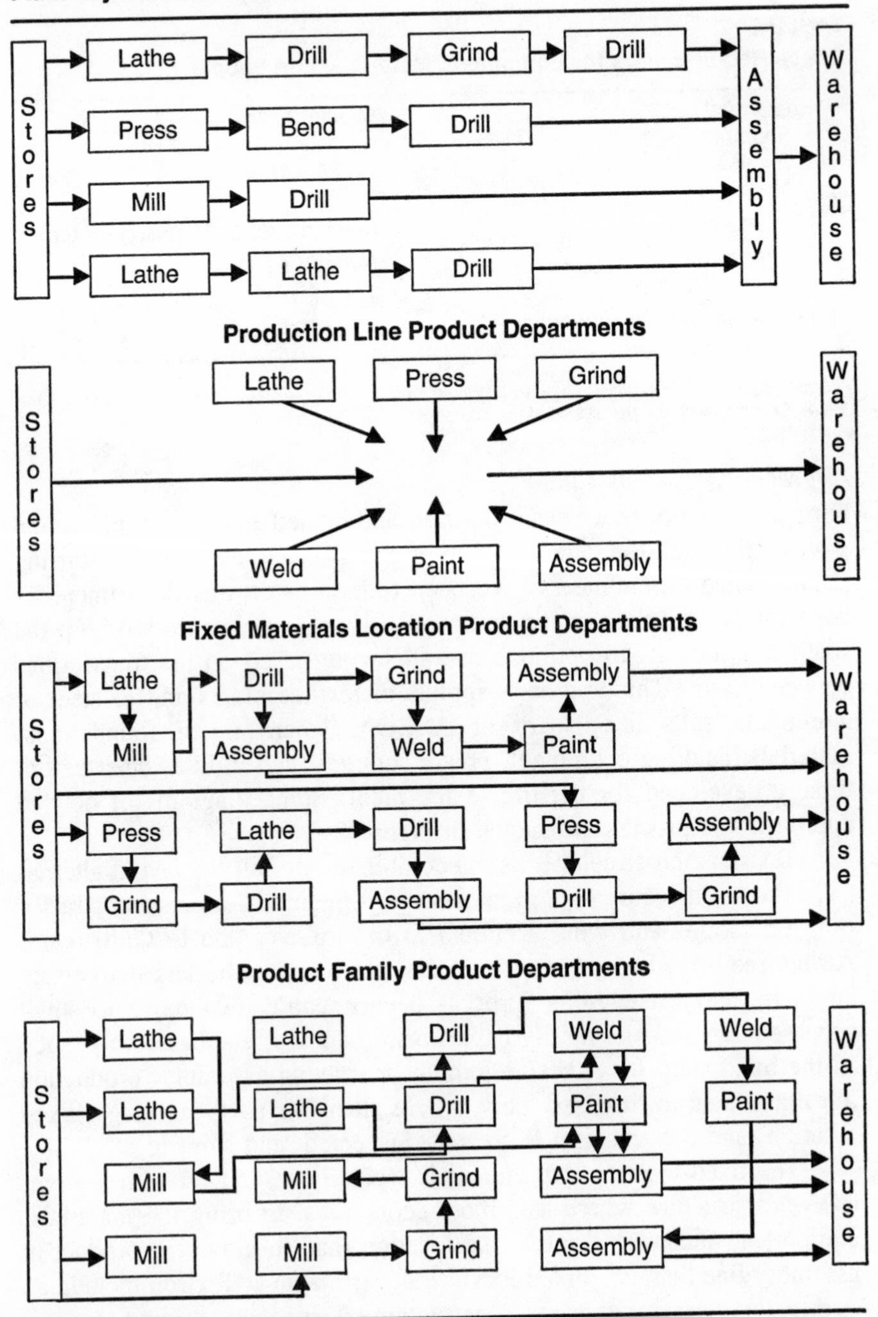

Source: James A. Tompkins and John A. White, *Facilities Planning* (New York: John Wiley & Sons, 1984), Fig. 7–1, p. 70. Reprinted with the permission of the publisher.

Nontraditional Examples

Shipbuilding

The traditional process of shipbuilding exemplifies a classic fixed-location layout and operation. Thus, one of the major innovations of post–World War II ship construction was the development of a production line at Avondale Shipyards in Gretna, Louisiana, to produce Navy frigates and destroyer escorts in a flowing fabrication and assembly sequence. The whole facility design and layout was integratively conceived and executed. The steel fabrication shops were relocated along massive steel tracks, and the hull was built in a rotating cradle so that all welding could be done down-hand, which was more efficient and produced better quality. Eventually, the ship was placed into a fixed position, finished up, and side-launched into the Mississippi.

Avondale recently carried out somewhat the same task for the steel structure for a major hydroelectric power generating plant. Steelwork was prefabricated in 110-ton sections. The sections were side-launched, just like the Navy ships decades before, and moved upriver by towboat to the generating location. The sections were then positioned on prepared foundations in the already-dammed river, and the concrete was poured around the power plant. There were no major cost savings, but the project remained on schedule until completion, with considerably less site work.

Manufactured Housing

Clayton Homes, headquartered near Knoxville, Tennessee, is certainly the most profitable of the builders and marketers of manufactured housing—a product designed for *factory* construction. Only movement to the site, construction of the foundation, and the installation of utilities are needed to ready a unit for occupancy. Although there are many other regional factory home builders, Clayton has emphasized quality, flexibility, and affordability, within a number of experimental communities of "permanent" manufactured housing.

Just like Avondale, Clayton has disavowed the traditional on-site construction systems, using the company's many regional facilities to improve cost performance and deliver greater value to the customer. Clayton operates a number of "factories" for housing production, and the manufactured output may be used as a business office or other industrial facility. Not long ago, Clayton sold a full *shipload* of manufactured housing units to Israel to help alleviate the immigrant-induced housing shortage.

Creativity

Creative thinking reflects the same break with tradition that many companies have undergone on the journey toward JIT. Sometimes called the *clean-sheet-of-paper* approach, it is what good SD teams need to develop in their contingency planning and preliminary design efforts.

Responsiveness

There are two fundamental structural means of dealing with future facilities changes: modularity and flexibility. Where timing is uncertain, modular design enables phased building expansion (follow-on modules). Flexible design sizes basic utility systems for their current and future capacity (power, water, heat and ventilation, drains, cutting-oil piping) and incorporates them into main trunks or spines. The taps or feeders bringing utilities to where they are needed are flexible and easily rerouted. Flexible facilities thus accommodate later changes in layouts or materials flow that are not known in sufficicnt detail to be part of the current greenfield or grayfield contract (see Table 6–3).

The numbers in Figure 6–2 relate to various functional activities and suggest how the needed expansion capability can be achieved. Module 3 might be a warehouse, for example, and the materials handling spine would be tied to it from and to all fabrication and assembly areas.

TABLE 6–3
Means of Achieving Expansion Modularity

Overall scheme
Main utilities trunks (power, ventilation, liquids drainage, waste removal)
Main passageways
Personnel
Mobile equipment
Materials
Add-on without disruption

Source: John M. Burnham and Ramachandran "Nat" Natarajan, *Manufacturing Processes,* Instructor Guide (Falls Church, Va.: APICS, 1992), Fig. 2–27, p. 2–68. Reprinted with the permission of the publisher.

FIGURE 6–2
Facility Expansion Alternatives

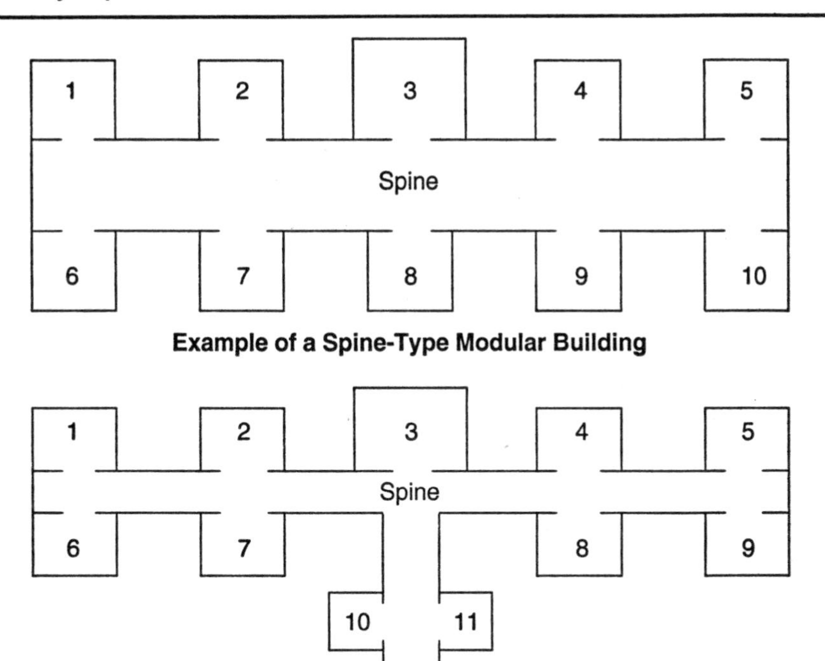

Example of a Spine-Type Modular Building

Example of a T-Shaped Spine Building

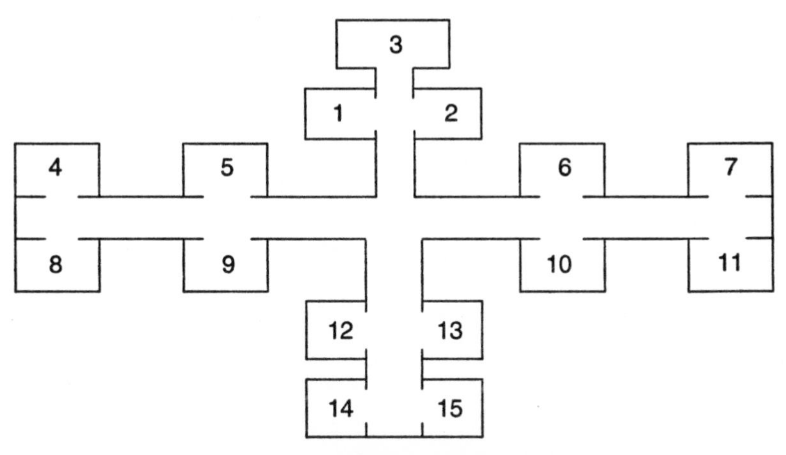

Example of an X-Shaped Spine Building

Source: James A. Tompkins and John A. White, *Facilities Planning* (New York: John Wiley & Sons, 1984), Figures 7–13, 7–14, and 7–15, pp. 248–49. Reprinted with the permission of the publisher.

Flexible design establishes main trunk ventilation and air conditioning, power mains (equipment and lighting), water, drains, sewage, compressed air, and lubricating and cutting-oil mains. These will be based on the fundamental plant mission.

Lacking definition of future details requires physical contingency planning: equipment sizing, power and ventilation subfeeds and slip joints, and flexible cables and hoses. These bring the needed utilities safely and easily to any location within the service zone. Similarly, main materials-handling passageways or trunks can be permanent, while individual feeder legs can be set up quickly with temporary fittings.

Thus, the modular approach minimizes the initial building investment, while enabling orderly expansion as growth takes place. Flexibility applies to workplace organization (layout, materials handling, and how people work) and the facilities support that will be required. This way the costs of grayfield or continuous improvement modifications will be minimized. Taken together, these two methods can lead to a site development plan reflecting everything that is known about conditions in the first phase of operations, and ensuring the LLCC over the plant's economic life.

Equipment Choices and Effects on Facilities

Not surprisingly, the same approaches—modularity and flexibility—can be applied to equipment for production and materials handling.

Here, modularity might mean multiple units of the same piece of equipment—contrasted with one larger, higher capacity machine, the latter perhaps more efficient but not able to accommodate mixed product batches easily. Modularity might also apply where initial product demand would not make good use of the big machine. And finally, if the manufacturing concept expects to use group technology and machining cells in the future, multiple units of smaller machines would be helpful to enable dedicated cell assignment later.

Equipment flexibility can be described in different terms: versatility, where the inherent design of the machine makes it able to do many different tasks, and adaptability, where the machine can be easily transformed to do other work. The historical practice in the United States has been toward versatility, and only recently has quick changeover capability been identified as a means of achieving flexibility. Note that with both properties, a single high-tech piece of equipment *may* be able to economically handle small lots of widely varying characteristics.

Equipment choices affect not only equipment capital costs but all the supporting systems too. Good contingency planning will anticipate process steps (or areas) where flexibility may be needed and make the appropriate facilities and utilities decisions to provide for it in the plan.

Manufacturing Software

Product variations within a family may lead to process variations—alternatives that can be exploited to yield effective equipment groupings, flow patterns, and logical locations for those work areas having intensive transactions. Decisions made during the facility design process can either facilitate or hinder the manufacturing task. Having an accurate, detailed product and process database is important both for facility planning and for the control of execution.

Layout

Layout is almost a given, once systems analyses are completed and equipment choices have been made. The materials flow needed to enforce the manufacturing concept and the related external flows to support manufacturing operations determine the layout, where this flow can be visualized. But in their absence (for example, in the production job shop), the classic department-by-department organization internally, and then the relationships and volumes/distances considerations among departments, are used to determine layout. A variety of classic layouts was shown in Figure 6–1. Depending on the systems requirements, modern plants are usually some combination of these layouts.

Details of Physical Organization

Facilities planning is an iterative process, constantly looping back and forth among the activities the plant must support and including the space and layout needed for them to be carried out effectively. Understanding these relationships requires various approaches using systems analysis: *organizational planning and control* systems and the manufacturing concept, *flow* systems for each factor of production, *environmental* systems, and *process-related* systems that describe special support requirements for the products and processes involved. Taking these separately, with relevant segments of each, and then integratively, will yield a systemic view of the facility's roles and how best to meet them.

It may be worthwhile to examine physical systems analysis while considering the approach used by Hewlett-Packard in designing and then implementing the campus concept, now almost an industry standard (see pages 68–70, and 91–92). Because of the product/process dynamics and corporate decisions on plant size and degree of independence, the sizing is essentially top down, and the facility details are developed to stay within established boundaries. Further, because it is impossible to stipulate what the essential departmental relationships will be in the future, the earlier discussion of flexibility in the facilities planning stage is especially relevant.

To use another familiar example, JIT manufacturing methods will lead to point-of-use storage, if any, and relatively small, decentralized "retail outlets" for parts peculiar, fasteners, and consumables. This contrasts sharply with the typical centralized (and locked!) storerooms of more traditional production. Note that organizational, control, and flow systems must each be involved in the facility definition. The activities and support due to special product or process characteristics will use specifics.

Materials Handling

> It has been estimated that between 20 percent and 50 percent of the total operating expenses within manufacturing are attributed to materials handling. Furthermore, it is generally agreed that effective facilities planning can reduce these costs by 10 percent to 30 percent. *Hence, if effective facilities planning were applied, the annual manufacturing productivity in the United States would increase approximately three times more than it has in any year in the past thirty-five years.**

Perhaps the most important reason why consideration of materials handling has been so neglected is that its costs are not usually measured. The operating costs of materials handling are classified as indirect, and so not scrutinized, unlike the more conspicuous direct labor elements. The capital expenditures for materials handling are usually lumped in with equipment purchases, construction, and start-up costs and are not tracked for effectiveness. Frequently, the most significant costs are not even those of running the physical system, but of WIP inventory and the need for computerized tracking and control. Beyond this, materials handling is taken for granted—a given in the manufacturing cost equation—and thus ignored.

---

*James A. Tompkins and John A. White, *Facilities Planning* (New York: John Wiley & Sons, 1984), p. 5.

Storage, Control, and Movement

These three traditional areas invite careful evaluation because of the no-value-added activity. Good analysis will address short move distances, the point-of-use storage concept, and only limited amounts of material.

Some *storage* systems use the same ideas to be found in good supermarkets: as long as items remain on the shelves, don't replace them. Have replacement inventory in the back rooms, but depend on overnight replenishment from the warehouse, not large volumes of stock just in case.

Control will follow from the kind of shop floor discipline that works with the manufacturing concept. If, as with supermarkets, it is a "pull" system, then controls can be visual for moderate fluctuations in demand rate. For forward planning, significant changes in production level or mix will require MRP gross requirements, along with good supplier coordination, to support the new production schedule. For the inside shop, there may be capacity adjustments that will also depend on the manufacturing concept and the flexibility of the work force and of the manufacturing processes.

Movement, defined by the routing for the production process, is optimized through systems analysis. If product line, flow manufacturing, or cells for different parts or assemblies reflect the manufacturing concept, then movement must follow. And if the production control system is self-regulating via pull, then materials movement is self-regulating too. Many JIT plants have gravity roller conveyors, sloped shelves to hold *a single* container of parts, the next piece of equipment close by, and relatively compact overall layouts.

Flexibility in materials-handling systems can be achieved through inherently adaptable systems: manual movement from the last work station of one part or one small container at a time; gravity conveyors, chutes, and slides; mixed-load pickups and distribution routings using independent wheeled vehicles or totes; or wire-guided automated tractors for hauling unit loads. Depending on product and process dynamics and the maturity of JIT/TQM efforts in the plant, contingency planning and systems analysis can suggest the most effective choices.

Computer-controlled storage, retrieval, and movement can also offer flexibility. Heavy-equipment manufacturers like Caterpillar and Ingersoll Rand and high-volume systems companies like IBM have used these AS/RS for many years to good effect, because the same systems, combined with MRP, could be used for both production line resupply and support of the replacement market demand for service parts. Movement from warehouse storage to point-of-use can be handled entirely by computer instructions to automated equipment, and facility-wide conveyor systems. Carousels can provide the same capability for small parts. Flexibility in computer-controlled systems is achieved by reprogramming and maintaining accurate location and quantity records. But the *physical* materials-handling system may be anything but flexible. Here other options must be explored.

Modularity in materials-handling systems design can yield some of the same benefits it does in effective building and utilities design. Main trunks and service spines can be installed, then extended with the addition of more production space. The "twigs" that bring materials directly to point-of-use can be light-duty and temporary in many cases, easily moved about to suit changes in the facility. This fits well with continuous improvement, though the logic is independently derived.

Key *integrative* questions arise in the process of materials-handling design, especially across boundaries. These may be departments in a single building, traffic among buildings, transport vehicles, and supplier or customer warehouses. These questions fall into the why-what-how category that must be applied to achieve integrative materials handling within the complex.

Logistics and Materials Handling

The goal of a logistics system is to gain competitive advantage through shorter lead times to the customer, but without premium freight, "JIT inventory," or enormous manufacturing penalties. The system is designed in terms of *unit load*— that quantity of a product to be packaged, palletized, or otherwise handled as a unit—generally an integer multiple or fraction of lot size in the factory. Careful choices for these controllable variables will be optimal in fulfilling system goals and, ultimately, consumer needs.

Range of Materials-Handling Concerns

The discussion thus far has addressed materials-handling designs to meet *site* requirements. There are a number of other dimensions:

Timing, production rates, and resupply frequency

Movement quantity and container size (or kit contents)

Work position and readiness for use

Sequencing of parts for mixed assembly or for manufacturing cells
Safety for personnel and security for item or part
Protection in storage and during movement to maintain quality

And, beyond all these, there is the effectiveness of the system in terms of its cost savings and the profit impact achieved.

Materials Handling Summary

As is true for layout, materials-handling requirements follow directly from complete systems analysis studies performed during detailed facilities planning. Because of shrinking product life cycles and the dynamics of product development and mix, the materials-handling system needs to be as simple and flexible as possible. And where significant investment is involved, the *systemic* effects, especially in logistics, must be considered. With strong trends toward smaller plants, outsourcing, and short product life cycles, the inflexibility of powered conveyor systems often overcomes their labor-saving advantages.

In plants where the decision to automate materials handling has been made without evaluating the system for value added, the result may be automatic guided-vehicle systems moving parts from one work center to another, massive conveyor systems, and complex computer control. Complex situations may demand a simulation of the required systemic materials movement and evaluation of alternatives to meet the needs. The investment funds saved through simulating first have been substantial.

Maintenance Planning

IFM also has the responsibility for developing a plant's maintenance program to provide equipment with *sustainable* capability. Maintenance is much more effective through incorporation in the facilities planning effort than reactively when carrying out operations. (A detailed treatment of maintenance planning, operations, and trends will appear in Chapter 8.)

Physical Asset Security

The classic model of a fenced area surrounding the entire facility, with only a few means of access, may still be a requirement for defense contractors, but it is not usual for many businesses—at least, not obtrusively so. Part of the IFM task is to help make the overall design a security-conscious one. To protect data, IFM will focus on the locations where sensitive information is stored in paper or electronic form. Secure design will try to restrict access during nonworking hours by locking doors, requiring code identifiers, and route denial—the closing off of work passageways at the end of regular hours. After-hours access is then possible only through manned areas where security personnel can verify identification and need-to-know.

In cooperation with plant accounting, IFM can evaluate high-risk areas for physical pilferage (product or equipment theft.) By examining both the normal and the theft-intent routes, it is often possible to prevent theft by simply not enabling access with transport or materials-handling equipment, because it is not possible to remove items of significant weight without bringing a car or truck onto the plant grounds. Moving equipment through personnel passageways is only possible if they're wide enough. The production process design can include tracking small, high-value items (e.g., microprocessors, IC chips, uranium pellets) for inventory control purposes and for detecting other discrepencies as well.

Warehousing and Distribution Planning

Receiving and storage, parts warehousing, and the storage of products in the distribution system call for another kind of IFM planning, though all of the essentials are the same. Good systems analysis will help by examining the flows required by both internal and external customers. Materials-handling studies will help determine basic commonalities (unit load, basic production quantity, basic out-of-plant movement quantity, current and expected rate of consumption by product/item/group) and seek to optimize the associated systems.

The same three aspects—storage, control, and movement—apply to warehousing and to production area materials handling. Various locational schemes for storage have been developed, depending—as with in-plant materials management—on the performance criteria. Fixed locations help with errorless storage and retrieval and visual control. Zone locations place the items of highest demand (traffic) nearest the staging point. Random locations minimize the amount of storage space required, but these need computer-supported records and possibly, automatic storage-and-retrieval equipment. So the charter and mission for the system are involved in the warehouse design.

Receiving areas or warehouses can be feeders to the production process. And distribution centers can perform a variety of value-adding operations to minimize inventory and meet customer service levels.

Examples

In an electronics receiving facility, electrical wire is delivered in bulk, then cut to length and bent to meet the assembly requirements.

In a large electrical engineering job shop, a CAD/CAM system uses circuit diagrams and their conversion in the computer to full scale, to drive the manufacturing wiring operations. Done at the point of storage for the wire coils, the color-coded wires are cut to length and tagged with precisely the right numbers, and then the correct terminals are attached. The cart with the wires is then moved to the switchboard wiring assembly operations area for installation. The result is very simple and error-free production, a paperless system, and effective use of engineering information to facilitate manufacturing.

Steel, received in coils by train or truckload, is slit to precise widths and delivered daily. Sheet, received in various metal thicknesses by the trainload at a steel warehouse, is sorted to match various customers' consumption mix and delivered to the point of use adjacent to the flame-cutting equipment.

Grain and agricultural chemicals are received by barge or railcar and then bagged for regional distribution nearer to the farms that the distribution centers serve.

Warehousing Systems

There is a mandate to develop cooperatively the interfacing aspects of warehousing systems. Supplier vehicles or company distribution transport must fit the truck docks, and internal materials handling must also mesh well. Independently conceived facility design to achieve LLCC may condemn both the company and its logistics partners to uneconomic levels of operating cost. Targets for design must, of course, include construction cost as well as outfitting. But ultimately, the issues of cost must also include time-related criteria, operations, and responsiveness to changing needs.

Once the external interfaces are developed, the flow paths within the facility must be mapped to suit the storage and activity logic. There are issues of size and scale, including initial and future peak loads, and issues of what the unit loads are or may become. And just as in manufacturing, there is the issue of suitability of the facility for the operators and maintenance personnel. Morale and effectiveness are at least as critical as in production systems, because of the connection with the outside world. And if production facility design and the materials-handling systems tend to get minimal attention, this is doubly true for warehouses. But various economic criteria—utilization of volume, of equipment, of people, and, ultimately, of capital—must be balanced against the clear need for quick, error-free access, ease of handling in and out, safety of product and personnel, and inherent responsiveness to change through flexibility and modularity.

Dynamics and Economics

Because of the requirement to interface with others outside the managed system, the receiving and distribution facilities managers must be aware of any changes planned for either supplier or customer technology. Such changes can affect the facility's ability to operate effectively.

Office Planning

Although there are some physical materials that can be "flowed" through the offices, these usually are representative of information and have value for that reason. The IFM analysis activities set forth in the context of manufacturing apply, too, in office design: nature, source, flows, volumes, predictability, timeliness, value-added aspects, other inputs, precedences, transformations, outputs, destination(s), and quality impact. Executives, managers, and office supervisors all have necessary, often frequent transactions among themselves and with their subordinates. An office design—in addition to meeting safety, light, ventilation, and space standards, must consider the needs for such transactions and for timeliness of output. Fortunately, most offices can be flexibly configured and easily reconfigured to suit efficiency and effectiveness criteria.

Much more critical are the work cells, with their need for frequent communication, some of it face to face. Order Entry and Customer Services personnel have a great deal in common and often handle inquiries randomly. Coordination and cross-checking is facilitated by being colocated. Accounting is usually grouped into general, receivables, payables, and, in relation to the manufacturing area, cost or product accounting. Engineering, Production Planning and Control, and other similar groups also have both a needed and a natural affinity for each other.

Personnel needs must be addressed appropriately: bathrooms, water fountains, health, child care, food services, storage for personal belongings, office supplies, office equipment, and work equipment (desks, phones, calculators, computers, and so forth).

Physical security must be a concern, for both personal and company items of outside market value. There is a developing trend to locate support staff near related-activity areas, which puts material planners/schedulers, unit managers, engineers, and cost accountants out on the factory floor. Some adjustment problems occur for staff without manufacturing experience.

LESSONS FOR THE INTEGRATED RESOURCES MANAGER

Industrial facilities planning is highly complex, but the benefits to the company are tremendous, including the development of a sustainable competitive advantage. Today considerable time and attention are paid to product-process design and to capable manufacturing processes. The Association for Plant Engineering is a long-established group for industrial facilities design and management. Since 1980, another professional society, the International Facilities Management Association (IFMA), with over 12,000 members, has addressed improving the design, execution, and maintenance of facilities, focusing on offices.

There remains considerable room for design improvement. The historical focus in industrial facilities design is on the products and processes and, more recently, on quality, flexibility, and cost of those processes. The areas requiring improvement are those associated with integration—development of the multifunctional, multilevel site development team is one means of doing this—and exploiting all of the internal and external networks and contacts the team brings with it to improve the design.

And while there are exceptions, the bulk of the attention is paid to the implementation of technology rather than the integration of the industrial facility into the manufacturing and distribution system. There is little emphasis on the design of the industrial facility for manufacturing. Rather, the issues are those of safety, access for the handicapped, environmental concerns, and hazardous-waste management. And while each of these is important, they do not bear on many of the serious design and integration issues raised by modern industrial facilities *as profit centers* and as constantly changing entities that support the overall strategy of the company.

Indeed, the manufacturing system *can* give companies a competitive edge. For this to happen, however, the focus must be on excellence, not only for production equipment functions but throughout the facility. This means that the plant mission and charter must reflect the full range of tasks assigned and be accompanied by the resources to carry out the assignments. It means that site development cannot be left to the engineers, because many of the decisions involve CIRM issues: trade-offs, uncertainties, and concept questions that cannot be answered only functionally.

The purpose of this chapter has been to outline the many interdependencies among the "separate" functions described as CIRM when considering the industrial or office facility in its totality. SD teams have learned to handle these dependencies effectively.

STANDARDS TO ASSIST WITH DETAILED FP DEVELOPMENT

There are a number of useful standards that have been developed, primarily by industrial engineers, to help with layout, space requirements, lighting standards, ventilation, aisleways, machinery clearances for access, maintenance, and operations. There are also regulatory minimums and maximums for egress and exit, safety protection requirements, a variety of airborne and waterborne substances, and for noise at working locations.

APPENDIX

This chapter ends with some thoughts to take away with you—somewhat like the checklist of information needs for facilities planning that appears at the end of Chapter 3.

Site Development Management. Effective IFM involves timetable articulation, based on the construction and outfitting contracts, and time-to-market issues for manufacturing. The project completion schedule should be coordinated through the SD team with all other areas of the company, including human resources development professionals (for recruiting and training), suppliers (for logistics and access), marketing and customers (for expectations of product delivery and for feedback).

IFM Supporting Role. There must be concurrent *operations* planning activities (with maintenance, production, quality, manufacturing engineering, and HRD), to assure that the facility can be an effective contributor through the details of the construction specifications.

Emergency Planning. Contingency planning for emergencies (health hazards, safety, fire, flood, explosion) is as important as operations planning and is, in some industries, dominant. Maintaining close external relationships with the community, public protection services, regulators, and suppliers (to temporarily halt materials flow, for example), is vital.

Systems Analysis. Systems analysis is needed throughout; emphasize materials flow, modularity, and flexibility as the plan becomes reality. One or more industrial or facilities engineering professionals can enhance SD team effectiveness.

Operations. By definition, industrial facilities management must plan, install, and maintain the environment to support manufacturing and to comply with regulations. The SD team can assure that the site's face to the public and to employees is the proper one. SD is involved in the externals of appearance—parking lots, traffic patterns, and so forth—as well as the internals.

Proactiveness. The SD team is proactive. Because the economic life of the facility is so much longer than almost any product life cycle, capital and product costs must be separated when targeting lowest facility life cycle cost. And the dominant criterion is to gain a sustainable competitive advantage, not to minimize current costs.

Ongoing Support. Continuous improvement applies to the industrial facility as well as to its processes. Mission and charter changes can redefine facility objectives, just as JIT and TQM can redirect manufacturing tasks. Existing health services and trends toward company-sponsored daycare suggest that in-plant facilities are a possibility.

The Mixed Facility. The increasing presence of support staff on the factory floor means that industrial facilities are becoming "mixed." Environment, safety, and work space design factors probably have changed forever. Cross-training and teamwork provide many promotion paths, more challenge, and more opportunities.

Customer Expectations. The manufacturing system must be properly located and configured to provide products meeting customers' expectations. Design processes and the equipment for manufacturing must be capable of making parts and products that meet all customer and regulatory requirements.

Cost Reduction. The foregoing has shown the IFM tasks to be complex, but the breadth of site development offers a significant opportunity for cost reduction and productivity improvement. Data given by Tompkins and White suggest that smoothly operating facilities can *triple* the bottom line.

Facilities Design and CIRM. Conflicts are opportunities, if resolved. The simplified "connections diagrams" of Figures 2–2 and 3–2, with additional concern areas sketched in, are shown as Figure 6–3.

FIGURE 6–3
Facilities Design and CIRM

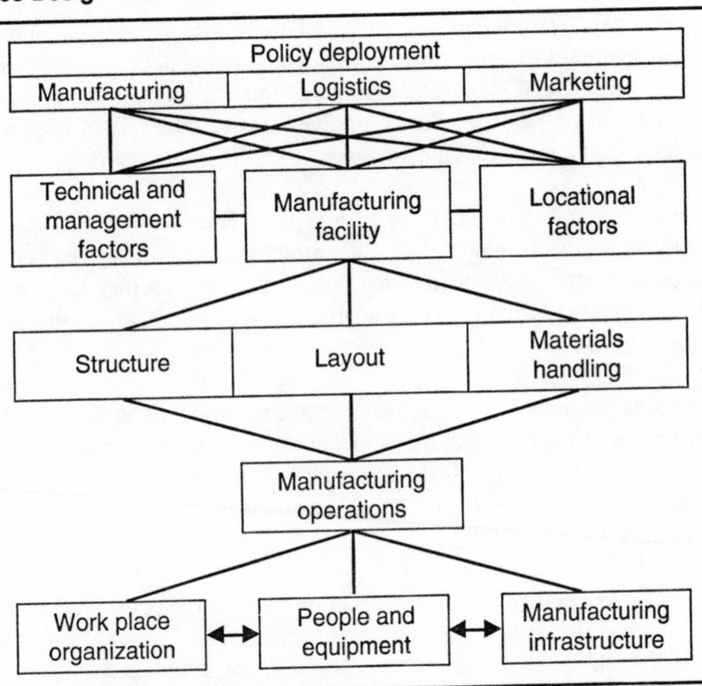

REFERENCES

Bell, Robert R.; and John M. Burnham. *Managing Productivity and Change*. Cincinnati: South-Western Publishing, 1991.

Burnham, John M. *Japanese Productivity: A Study Mission Report*. Falls Church, Va.: APICS, 1983.

Hayes, William H.; and Steven C. Wheelwright. *Restoring Our Competitive Edge: Competing through Manufacturing*. New York: John Wiley & Sons, 1984.

Heizer, Jay; and Barry Render. *Production and Operations Management*. 3rd ed. Boston: Allyn & Bacon, 1991

James, Robert W.; and Paul A. Alcorn. *A Guide to Facilities Planning*. Englewood Cliffs, N.J.: Prentice-Hall, 1991.

Skinner, Wickham A. "The Focus Factory." *Harvard Business Review* (1974).

Herman Miller Corp. *The Integrated Facility System: Production Work Stations*, Zeeland, Mich.: Herman Miller Corp., 1980.

Tompkins, James A.; and John A. White. *Facilities Planning*. New York: John Wiley & Sons, 1984.

CHAPTER 7

PROJECT MANAGEMENT: IMPLEMENTING THE FACILITIES PLAN

With a contractor selected, the execution of the facility plan begins. Architects, contractors, and the site development team, expanded to include many more specialists and prospective plant management and operating staff, undertake the challenging process of managing the construction and activation of the new facility. For any major project, this stage may take one or more years and involve millions of dollars and many thousands of man-hours. Teamwork will lead to a successful conclusion.

THE MUTUAL EFFORT

Project management is the means by which the facility plan becomes reality. Robert D. Gilbreath, in *Winning at Project Management,* calls it "a *mutual effort*, using a collection of resources in an orchestrated way to achieve a joint goal." And Peter W. G. Morris, in Cleland and King's *Project Management Handbook,* says:

> The most pervasive intellectual tradition to project management, whether in organization, planning, control, or other aspects, is without doubt the systems approach. A system is an assemblage of people, things, information, or other attributes, grouped together according to a particular system 'objective' . . .. Properly organized and managed, the overall system [and its related subsystems] acts in a way that is greater than the sum of its parts.

The site development (SD) team, augmented by various functional specialists, takes on the role of carrying out the plan through civil engineering layout, bricks and mortar, equipment selection, and the multitude of tasks needed to create the new greenfield plant or modify the existing one. For significant projects, most assignments are full-time, and the project team focuses entirely on the information exchange, planning, and controlling of the many subprojects that make up the end product: the plant.

Greenfield developments are *not* the only undertakings that justify detailed project management. A major process plant grayfield project can involve full-time staff for a number of years and expend hundreds of millions of dollars.

For instance, when ALCOA undertook its recent modernization of the North Plant of its Tennessee operations, grayfield planning began in 1983 with an expected 10-year program to completion. Through 1988, 4 million man-hours of training had been completed, with a continuing target of 20 days per employee. All major programs and projects were given full-time staff. The new continuous cold mill was delivered in 1987. Installation and activation took another year. The building of a new ingot facility was a separate but related project. Another contract addressed the materials-handling system for the mill. The hot line was completely rebuilt. All told, ALCOA invested nearly $500 million to guarantee world-class status for the Tennessee mill. Strong project management tools were employed from the outset—first for training, then for the actual construction and commissioning (see Table 7–1).

What goes into project management? First, a common understanding of what the various phases of the project (e.g., facility plan) entail, and

TABLE 7–1
Rules for Managing Projects

1. Set a clear project goal.
2. Determine the project's objectives.
3. Establish checkpoints, activities, relationships, and time estimates.
4. Draw a picture of the project schedule.
5. Direct people individually and as a project team.
6. Reinforce the commitment and excitement of the project team.
7. Keep everyone connected with the project informed.
8. Build agreements that vitalize team members.
9. Empower yourself and others on the project team.
10. Encourage risk taking and creativity.

Source: W. Alan Randolph and Barry Z. Posner, *Effective Project Planning and Management: Getting the Job Done* (Englewood Cliffs, N.J.: Prentice-Hall, 1988). Reprinted with the permission of the publisher.

the development within the augmented team of a common language for communications. Randolph and Posner call this *information* management. Second is detailing the execution of the facility plan. This is clearly not the same planning discussed in Chapters 4 and 6, but rather the how-to's of execution, after content has been decided. Generally, this project plan becomes the control system for evaluating performance—task times and budgets against work actually completed, for example. And finally, there is the program for execution, when resources—money, materials, equipment, and so forth—are committed to the job.

All management is essentially the same: planning, organizing, and controlling. But when addressing a major grayfield or greenfield facility construction and commissioning project, a tremendous amount of detail is involved, with a potential for it to become almost *un*manageable!

Enter network planning methods—today almost synonymous with project management. This offers an organized means of planning and controlling at various levels of detail. Program evaluation and review techniques (PERT) and the critical-path method (CPM) were developed to improve management of truly major undertakings—the Polaris missile–carrying nuclear submarine or a major passenger ship, for example. Although the early efforts were manual-visual networks and hard to keep updated, today's PERT/CPM programs are computer based and relatively easily maintained. (We assume that the details of PERT/CPM are known to most readers, so these will not be discussed further here.)

Site Development Team (Again!)

Just as concurrent engineering makes the time-to-market much shorter and the products better, so the blendings of the core SD team with a succession of specialists leads to a better physical and operational facility. The final site development team augmentation takes place when the construction project planning begins, adding more operational, results-oriented members to take on subtasks in support of the project completion timetable.

Formal phases that relate to the same kind of time-based hierarchy have been successfully applied in major projects, like the Apollo space program. These are described as the life cycle management approach and are quite literal in terms of the titles of each phase: conceptual, definitional, production or acquisition, operational, and divestment (or post-completion). Some details appear in Table 7–2.

TABLE 7–2
Project Life Cycle Management

Conceptual Phase	Definition Phase	Production Phase	Operational Phase	Divestiture
What's the problem?	*Getting serious*	*Project management operations*	*Putting the product to work*	*Transfer it to operational users*
Needs	Detailed resource needs	Operational planning	Operations	Permanently phase out project team
Applicable concepts	Support plans	Obtaining resources	Integration into "normal" organizational systems	Plan and execute transfer to operations
Feasibility	Cost-schedule-performance requirements	Validating specifications and timetables	Multi-dimensional evaluations	Include post-project evaluations
Tentative solution	Contingency and risk management	Begin operations	Learning feedback for the future	Complete the support and technical documentation
Resource requirements	Interfaces studies	Work breakdown and assignments	Support systems appraisal	Develop appropriate analyses for future project managers
System design	Specific support needs	Testing and validation		Technical researchers, and contractors
Project organization	Initial and ongoing information	Documentation (ongoing)		
	Policy, and control documentation (budgets, documents, job descriptions . . .)	Planning for support during operations		

Source: Adapted from David I. Cleland and William R. King. *Project Management Handbook,* 2nd ed (New York: Van Nostrand Reinhold, 1988), pp. 194–196. Reprinted with the permission of the publisher.

Levels of Project Management

Just as there are natural hierarchies based on different planning horizons, within an industrial organization (see Table 1–1), hierarchies in project management are for the most part defined by horizon length, level of detail, and the makeup of the constituency, or stakeholder group. Morris suggests that senior management must relate with the world outside: Congress, the executive branch of the federal government, and various regulatory bodies. This parallels some of IFM's *representing* responsibilities in the corporation. There is a distinct coordinative *(enabling)* role for middle management for the technology and a protective role to buffer projects from outside disturbance.

Project Planning

Specific objectives are the driving force toward goals. For example, one can translate GM's "competitive compact car goal" into a set of objectives, such as "Saturn New Product Launch in Showrooms Nationwide by September 1989." As noted earlier, senior management must negotiate these goals with full knowledge of the expectations (or even the mandates) of various stakeholders, expectations which may differ with the nature of a person, an organization, or a legislative body. Earlier representation in these policy discussions by the project team leadership can enhance the results once resources are committed.

At the other end of the spectrum are project tactics—those detailed plans, checkpoints, activities, relationships, and time estimates used at the subproject level to accomplish and maintain control over the assigned tasks which *must* mesh together (the current language is that of a "seamless transition") to support the related objective(s).

Contingency plans were discussed in Chapter 4 as a means of preparing for uncertain events, and have somewhat the same context here. The objective of project contingency planning is to anticipate deviations (task time, budget, quality, or precedent event failure in this regard) and to think through how best to recover. Note that the "Saturn launch date" objective was changed several times, based on additional information gained through actually carrying out early phases of the project.

In PERT/CPM terms, the completion times of various tasks can be shaped by expending more resources or "slipping" some task beginnings, endings, or durations. By doing some contingency planning, however, the project engineering team can choose the most effective cost and time activities to focus on for schedule restoration.

The use of the project (task) plan as a control device brings up another consideration: Is there a problem with *execution* such that the remedy may lie there? Or is there a difficulty with the plan itself (e.g., it's unrealistic, missing details, or has errors) that requires correction before determining changes in current or future activities? Both are clearly possible, and contingency planning needs to consider the most effective revisions to both.

Performance Factors

It should be evident that effective project management cannot take place without a plan and appropriate control factors and measurements. At the same time, various levels of project management are confronted with the potential of a moving target. One circumstance is the changing benchmark: what must be achieved to be the best. The other is related to the internals: product and process dynamics which may evolve even as the facility is being built.

As an example of the many required interactions, consider Corning Glass Company. When Corning first developed laboratory processes to produce the multiple glass fiber bundles called *optical waveguide*, plans for a new manufacturing facility were set in motion. It was Corning practice to learn by doing and carry laboratory practice into production to work through the bugs.

But the facility, once built, sat for three years without orders, because Corning engineers continued to develop better ways—better equipment and processes—changing the untried plant again and again. Eventually, the company ascertained that the cost, quality, and production capabilities were appropriate to the market's demands and thereby set the benchmark for world production for optical waveguide. (For more insight into this fascinating project, see the detailed description in Magaziner and Patinkin's *The Silent War.*)

Tools for Project Management

The varying memberships in a site development team require strong technical and human skills on the part of the managers involved.

A "matrix" organization allows team members to retain identity with the specialty or functional area they have been working with and to maintain that connection for administrative purposes. But on either an ad hoc or formal basis, project team members come under the leadership of the project engineer or manager and respond to assignments as team members. The project manager carries out the liaison with the functional managers who have lent people to the project, reporting on job progress, the duration of the need for that person's support, and on team member performance as a basis for the individual's ratings and merit recommendations by the functional department manager.

Projects depend for success on networks, human contacts, and interpersonal skills. Also there needs to be quite formal planning, especially for large, long-duration, major-resource-consuming programs. This has been noted in almost all government and aerospace projects and in the commercial sector—for capital-intensive businesses like steel, refining, and paper making. As these industries carry out capacity expansion strategies, such programs will be centralized (at the corporate level or equivalent) both before and after the facility construction and equipment installation phases, but decentralized (site development) during implementation.

Getting Started

Sketching the Network

Goals and objectives drive the program. With these established and with the appropriate level of team gathered, the next step is to rough out the schedule. Either Gantt or network modeling will be helpful to convert concept into a communications device. The schedule becomes the centerpiece of the team management effort. Before it becomes finalized, there will be many revisions and improvements, each leading to a clearer understanding of what lies ahead for the team—including, perhaps, its augmentation needs.

Manageability

The overall program is too complicated to understand *in detail*. Breaking it down into smaller portions is essential. These subprojects, sometimes called *work breakdown structure,* become logical assignments for subteams to develop and, later, to manage. Those familiar with manufacturing can see immediately the analogy with product, process, and part design and the management of all parts to make a shippable end item. Conversion from ideas and concepts into actual activity details can both plan and control the subproject work itself.

Schedule

Several related project-planning activities must occur at any level of concept-definition-acquisition planning: the determination of *precedence*, the estimates of *time* required, and the *cost* of accomplishment. Following comes the planning for *control* purposes—to show progress and to compare results with budget and planned accomplishment on a time-phased basis.

Projects, as Moder (1983) puts it, are "one-time efforts . . . planned and scheduled on the basis of experience with similar projects applying PM judgment to the particular conditions of the project at hand," and the task of project management is that of directing resources over time to accomplish the stated goal(s).

FIGURE 7–1
Work Breakdown Structure Information Integration

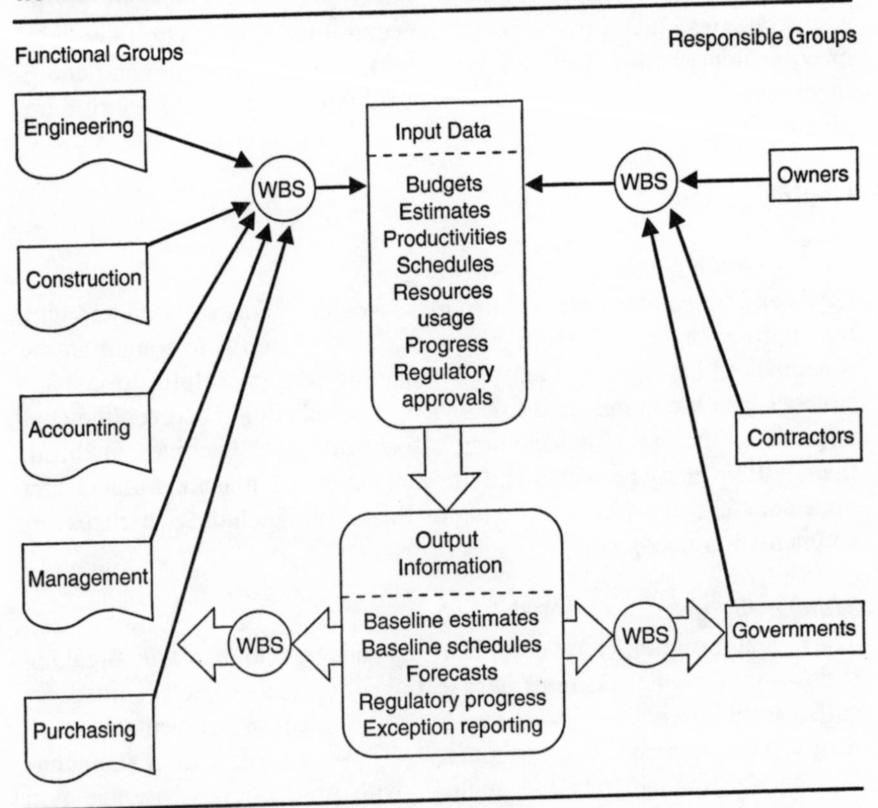

Source: David I. Cleland and William R. King. *Project Management Handbook* 2nd ed. (New York: Van Nostrand Reinhold, 1988), p. 304. Reprinted with the permission of the publisher.

The *sequence* of activities and the time and resource budgets to complete them must be worked through and included in the plan. The assemblage of all such activities related to a particular subproject or work program allows both consistent diagramming and a prediction of the duration of the subproject as a whole (see Figure 7–2).

Small modifications with tight timetables—modifications requiring plant shutdown, for example—also benefit from good project control techniques. Most managers feel that the discipline of doing the planning using either PERT or CPM is a primary benefit. Once the interdependencies are clear, the project can be risk-managed to completion on time. Lacking the perspective and the planning estimates that a good plan generates, the manager can only react to circumstances as they occur.

Elements of Preparation for Project Management

Before launch of the site construction program, the various aspects of risk assessment and contingency planning, examination of possible unfavorable outcomes, and the availability of corporate or division level policies and standards that can reduce consideration of various eventualities are all factored into the execution and controls system that will govern project activities.

Risk Assessment

Knowing the criteria by which the SD project success will be judged can have a great influence on how the project will be managed. For instance, if product announcements and commitments to marketing have been made, then completion of the greenfield or grayfield project *on time* will have higher priority than staying strictly within budget.

Budgetary Factors

Meeting limits of total-project or major-activity cost may be dominant. If recession has reduced demand, then fiscal considerations may actually suggest *reducing or deferring effort* or even going to a hold status. Knowing how to accomplish this is very important. The abandonment of the prospective TVA Hartsville nuclear plant is a case in point.

Quality or Technical Issues

If new technology obsoletes processes already installed (as in the case of Corning and the optical waveguide), it's necessary to "go back to the

FIGURE 7–2
Project Management Using PERT/CPM

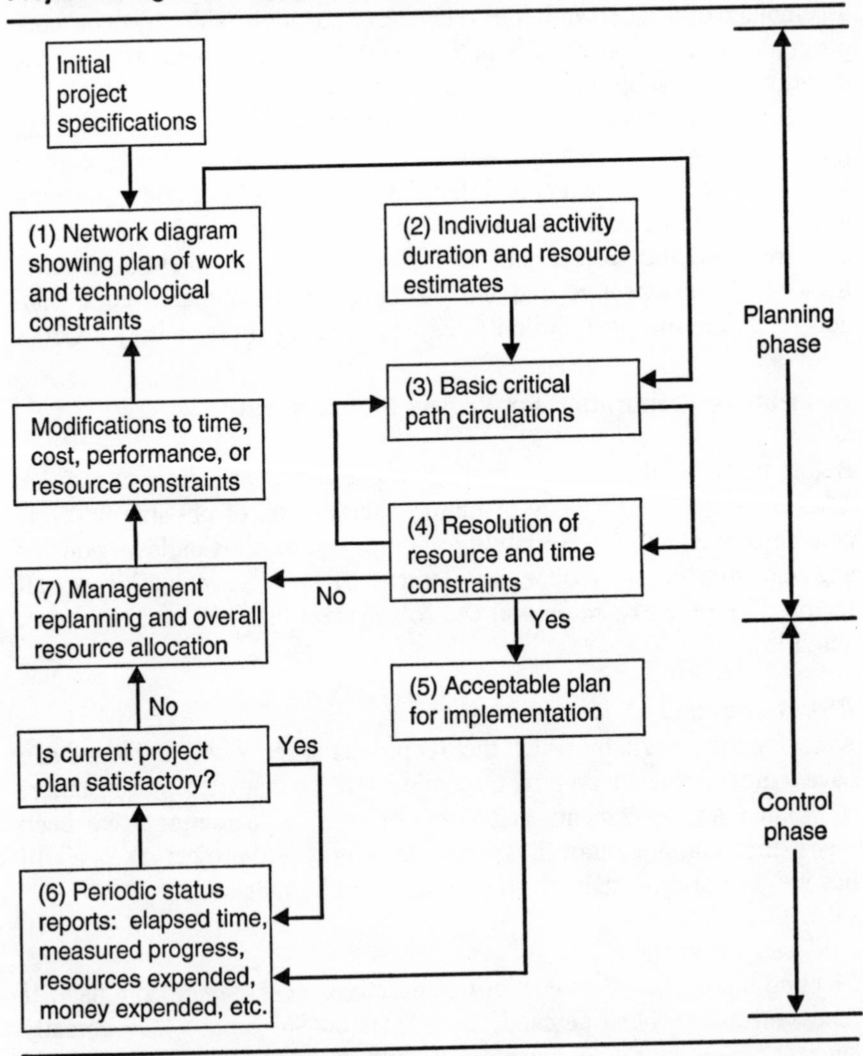

Source: Joseph J. Moder, "Project Management via PERT/CPM," in David I. Cleland and William R. King, *Project Management Handbook*, 2nd ed. (New York: Van Nostrand Reinhold, 1988), Fig. 15–2. Reprinted with the permission of the publisher.

drawing board,'' but with some sense of how to achieve whatever economic recovery is possible. Issues of product quality may be very important. The GM/Saturn launch was delayed for quality reasons, affecting many aspects of the facility development. Aerospace safety was always critical, and it became even more so following the Apollo Three and Challenger tragedies.

Trade-offs

Trade-offs are best considered, and their impacts assessed, before a crisis arises. Computer packages are particularly helpful to such analysis because, with so many possible combinations of variables, manual solutions might miss the best opportunity. This applies to finding new solutions that "crash" the project to catch up on schedule slippages or that reduce spending rates (or reallocate them) to stay close to the original completion target, but within the available funds.

Availability of Standards

Randolph and Posner provide a list of the types of standards of practice (standard operating procedures) that many companies maintain to help project teams focus on the unique aspects of their work rather than those already stipulated by the SOPs. The list of these standards, shown as Table 7–3, is suggestive of what probably is available within your own organization. To these could be added the construction and amenities standards (e.g., of Hewlett-Packard), the environmental compliance standards, and other mandated stances taken by a company in its dealings with regulators. Insurance and emergency planning considerations are also likely to have the effect of regulation, though not always administered by *public* authority.

PROJECT MANAGEMENT ACTIVITIES

With the planning and risk assessment phases and preparations completed, the operational aspects of project management take over. Contracts for building design and construction, purchase and installation of equipment, utilities, site preparation, parking lots, roadways, and so forth have been either established or planned for. The site development team has been closely involved in all of these activities, and individuals or groups track the current status of planned activities or subprojects, divided along lines of interest and special competencies. What happens now?

TABLE 7–3
Types of Standards

Policy	*Specifications*	*Processes*
Procedures	Designs	Responsibilities
Organizations	Change categories	Budget levels
Staffing levels	Procurement types	Audit types
Funding criteria	Quality programs	Research steps
Plans	Written copy	Personnel management
Graphics	Scheduling levels	Pricing methods
Contracts	Materials	Accounting codes
Equipment types	Proposals, bids	Information systems
Reports	Configurations	Work breakdowns

Source: Alan W. Randolph and Barry Z. Posner. *Effective Project Planning and Management: Getting the Job Done* (Englewood Cliffs, N.J.: Prentice-Hall, 1988, p. 299). Reprinted with the permission of the publisher.

The SD Team

First, some suppositions. The project managers have achieved buy-in among the members of their group and have ratified their common purpose. The SD team members have been empowered to act on behalf of the project and feel responsible and accountable for what takes place. Communications have been well established among all the "core" and specialist members, and a common project language exists or has been developed to facilitate understanding.

Communications

Focus begins with communications. Within the team, all parties must "feed the kitty" to keep project data available and fresh. Knowledge is power, and the team must, like the army ant colony, share a collective intelligence about all relevant aspects of the project "terrain."

Project progress reports must be available, in the communications style and with the degree of detail that meet the requirements of the customer. Regular reports can use the control systems established to manage project activities, and electronic data interchange can make transfer of relevant information both timely and inexpensive. But remember: An uncommunicative project team can lose credibility at headquarters despite excellent accomplishments at the site.

Networking, always an important informal communications device, is very helpful in maintaining a proactive position while focusing on operational issues. Learning what is happening elsewhere in the company links team members and their functional specialties, assures temporary help if needed, and collects grapevine material that may help anticipate changes affecting the project.

The benefit of planning is that the team will only have to address a limited number of absolutely new issues—a kind of management-by-exception principle. Following the discipline that already has incorporated all known information into the facility plan and addressed configuration uncertainties through modularity and flexibility, the team must address new issues in the same way. While all team members have been exercising their individual empowerments to deal with assignments according to plan, the team must gather to jointly address and conquer the new problems that arise. Focusing on objectives and goals, but not limiting the means and tactics for their achievement, can lead to better solutions than those that might have been planned!

Technology

Changes to product, process, equipment, materials handling, layout, procedures, manufacturing concept, or regulations affecting product or by-product handling or disposal can occur between planning and execution. Construction code changes may disrupt structure or access, and adjustments due to technology will definitely occur. The team, augmented by temporary members who can help with impact assessment, must think through the problem and minimize the negative effects on project success.

Control

Control obtains feedback on system activity, compares it with the desired outcome, and provides information to management. This holds true for project systems as well as manufacturing. If the PERT/CPM diagram of the project represents the plan, then it seems reasonable to use it for both progress reporting and for control. To do this, the PERT plan is periodically rerun to reflect actual status for time (and cost, if appropriate). Because we can affect the rate of progress only on the work remaining, we can evaluate alternative ways to get back on track, given knowledge of time-cost trade-offs.

Figure 7–3 illustrates J. J. Moder's project cost versus time curve. It was developed by using the original plan and then updating it to reflect the resources expended, the progress made, and the time remaining. The results to date are extrapolated to allow estimates of the total resources, the total time that will be spent, and (in this case) the overruns.

Throwing money at a problem is not the only, or necessarily the best, solution, nor is maintaining the same expenditure rate and accepting the results. As noted earlier, the criteria for success may not be just time and cost. Some analogies with more traditional shop floor manufacturing practices exist: alternate routings, lot splitting, overlapping, overtime, and subcontracting are tactics that can be used with projects as well as with manufacturing.

Adding extra personnel may help. The usefulness of this approach depends on the critical path and how sensitive *other* activities may be to

FIGURE 7–3
Project Cost versus Time

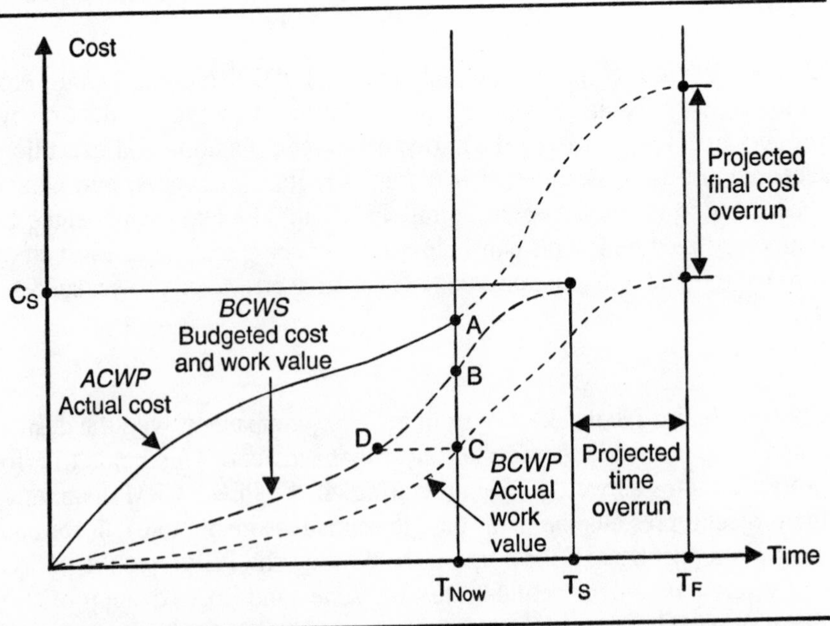

Source: Joseph J. Moder, "Project Management via PERT/CPM," in David I. Cleland and William R. King. *Project Management Handbook,* 2nd ed. (New York: Van Nostrand Reinhold, 1988). Reprinted with the permission of the publisher.

changes in any particular one. Network analysis can provide this information. Sometimes other activities can actually be slowed down (*new* earliest versus latest start date) and the effort applied where it is needed most. PERT implies only sequential, head-to-tail activities with all predecessor events completed. This may not be true, and some recovery can be made through deliberately paralleling or overlapping activities that were defined as serial. As with production lots, sometimes part of an activity is critical for starting others, and the remainder can be done later—a sort of lot-splitting concept. This is where the team's creativity and intimate knowledge of the job come into play.

Cost Control

Moder's diagram (Figure 7–3) provides a cumulative measure of cost versus progress. It can be helpful to accumulate costs by work package, or bid line item. The work structure breakdown (Figure 7–1) can also be a useful tool. Subproject engineers will be concerned with their own project and others on which it depends, while middle management may need the viewpoint of Moder's curve. At the program level (or in the customer shop), there may be overriding concerns like those of GM/Saturn and Corning that negate traditional control techniques.

Precontrol

The plan-to-actual comparisons are much like inspection of the finished product: Good parts can be separated from bad ones, but the quality of the process is not improved. This suggests that sampling must give way to mistake-proofing and, ultimately, to design that will preclude off-quality production. The contingency planning that has been referred to throughout this book has the goal of anticipating difficulties and preparing to deal with them—a sort of precontrol. Similarly, each project has specific risks known at the outset. Those affecting primary project success criteria will need sustained attention.

Legal and Regulatory Concerns

Safety at the worksite; disposing of offal and scrap; maintaining insurance coverage for the company, contractors, and visitors; dealing with medical or other emergencies that can arise—all these must be managed as a part of the project effort. There may be local requirements for vehicle movement, air quality, security, and noise. And construction site environmental and workforce safety regulations abound at the local, state, and federal levels. Contractual provisions must be met. And there are significant security needs as factory hardware and equipment moves onto a construction site.

PROJECT CONCERNS

Project management is still *management*, and it benefits from adherence to basic principles. An empowered SD team can't operate with arbitrary, top-down decision making, nor can the team succeed without sufficient authority and influence to get the work done. Conflicts within the team must be addressed and resolved. Communications with interested parties must be timely and in appropriate detail for the consumer. Contractor personnel are part of the execution team, not enemies. New technology may be hard to integrate because of unfamiliarity. Rule changes or an unfavorable community, customer, or supplier environment can call for prompt regrouping and refocus.

Baker et al. have provided a list of management characteristics that strongly affect the perceived success or failure of a project. The most statistically significant factors are the following, in order of their explanatory power:

1. Coordination skills and relationships.
2. Adequacy of project structure and control.
3. Project uniqueness, importance, and public exposure.
4. Success criteria salience and consensus.
5. Competitive and budgetary pressure.
6. Initial overoptimism or conceptual difficulty.
7. Internal capabilities build-up.

Baker et al. note that the fifth and sixth factors, if present, have a significant negative impact.

COMMISSIONING AND START-UP

"Cleaning up" a project involves a great amount of detailed work, and the temporary addition of experienced people can help. The continuous observation of all aspects of the construction, finishing, and outfitting of the facility results in the correction of many defective items before "punch list" time. The people who will be operating and maintaining equipment can help greatly, and they may be "enlisted" as part of the augmented SD team during the late phases of construction and for most of the installation and checkout activities. The project team members and representatives of contractors and operations (manufacturing, office staff, and so on) will finally develop the exception lists that become the basis for acceptance ("beneficial occupancy") and a move toward operations.

When, section by section, the facility is accepted, the contractor locks personnel spaces and turns them over to operations. In manufacturing, the crossing out of exceptions and the final removal of contractor equipment and personnel from a department or area can become a basis for effective completion.

There are any number of legal ramifications of how *not* to manage this process. Project team members and entities like contracting officers or owner's representatives must have a common understanding of individual and shared responsibilities. The effect of their activities and approvals is to move the contractual relationships from performance as a development of the plans and specifications into the mode of discovery and correction of hidden defects—equivalent to the warranty provisions of most commercial contracts.

Once the operating personnel accept responsibility for the facility, the formal commissioning can take place, checkout and dry runs take place, and the facility moves toward production status. Trial batches of materials are ordered; machine power and cooling and lubricating systems are activated; personnel training and checkout take place; and various performance tests are designed and carried out. Generally, these have a logical bottom-up sequence: from support systems to individual pieces of equipment, to subsets, linked machine cells, transfer lines, or other interdependent process steps, and then to final assembly, packing, and shipping. The role of in-line inspection or testing will be integral with the processes being tested, and issues such as capacity, capability, and reliability must be examined from the operating standpoint.

For a challenge, consider the design and construction of a merchant ship. The construction and on-shore test phases are similar to those of any other industrial or office facility. The final acceptance process is called the sea trial, which involves several days of standardized performance tests of the entire ship system. Because the ship has unique self-support requirements—including those of stability, efficiency, and maneuverability—tests of those capabilities are carried out at sea. Water ballast is substituted for cargo and passengers, and the ship is put through her paces in simulated operating conditions. Vibration throughout the ship provides the opportunity to hear loose clamps or screws, cracked welds, misalignments, and so forth. It's an exciting but exhausting time for all concerned. And when all the performance tests have been completed, the trial board and most of the temporary project staff disband and return to other assignments, while the master and chief engineer go about the business of readying the ship for work.

LESSONS FOR THE INTEGRATED RESOURCES MANAGER

There is a philosophy, a logic, and a set of well-established techniques to help managers successfully carry out projects. Understanding and communicating the nature of the task and the goals established for its accomplishment are critical. Building the team to include the needed experience and skills, and then helping it to become like the army ant colony with shared intelligence and a common purpose, is also critical.

Adequate resources, the authority to augment the team, and the power of accurate current information about all aspects of the project will enable good project performance. The ultimate tests of project success are not just "on time, in spec, and on budget," although these are important; it's the degree of satisfaction experienced by the customer, the operators, and the project team. This is why understanding *all* the important criteria for evaluation is critical.

GM's Saturn needed to have a highly competitive vehicle and targeted Honda as the one to beat. Corning researchers demanded of themselves that the optical waveguide not only be feasible but establish an entirely new level of communications expectations in its customers. Before the project personnel on either of these teams were willing to turn the product loose, they had to be satisfied that it met all *their* expectations too. Technical performance—whether lighting or automobile or waveguide or ocean-going ship—is integrally tied in with the perceived project success, while time and cost are generally less directly related.

These facts suggest the following propositions for managers:

- The strategic goals that created the need for the project are important to the team, both technically and in terms of their motivation value.

- The right people may not be as important as the right spirit and the strong commitment that follows. Teams establish and manage themselves if buy-in happens and they feel empowered to act to achieve team goals.
- Control tools can greatly advance contingency planning and developing corrective actions. But PERT/CPM and Work Breakdown and Budget should not be allowed to dominate how the project success is judged by either insiders or outsiders.
- Communications and networking are vital at all stages of the project.
- Whatever the obstacles, effective coordination leads to group satisfaction and to a feeling of success.

REFERENCES

Cleland, David I.; and William R. King. *Project Management Handbook,* 2nd ed. New York: Van Nostrand Reinhold, 1988.

Gilbreath, Robert D. *Winning at Project Management: What Works, What Fails, and Why.* New York: John Wiley & Sons, 1986.

Kertzner, Harold. *Project Management: A Systems Approach to Planning, Scheduling, and Controlling.* 2nd ed. New York: Van Nostrand Reinhold, 1984.

Magaziner, Ira; and Joel Patinkin. *The Silent War.* New York: Random House, 1989.

Moder, Joseph J.; C. R. Phillips; and E. W. Davis. *Project Management with CPM, PERT, and Precedence Programming.* 3rd ed. New York: Van Nostrand Reinhold, 1983.

Randolph, W. Alan; and Barry Z. Posner. *Effective Project Planning and Management: Getting the Job Done.* Englewood Cliffs, N.J.: Prentice-Hall, 1988.

SECTION 3

OPERATIONS AND SUPPORT

CHAPTER 8

MAINTENANCE: PLANNING, OPERATIONS, AND TRENDS

For the facility to be an operational success, the productive equipment must be skillfully maintained, so that the processes remain capable of delivering the highest quality of products. Proactive maintenance planning can be integrated with developed operators' skills to provide sustained plant capability with little or no "lost time" and no scrap. Only through such proactive maintenance practices involving all of the workforce can Total Productive Maintenance be achieved.

THE MAINTENANCE IMPERATIVE

> Among the chartered functions of a facility management department, what stands out in most of the cases as the number one priority mission is *to eliminate or at least minimize production stoppages caused by facility breakdowns or malfunctions.** (italics supplied)

When delivered and newly installed, equipment has operating characteristics guaranteed by the supplier. An effective, proactive maintenance program can ensure that equipment has *sustainable* capability. The maintenance planning and operating effort is often coordinated closely with quality personnel because their analysis of tolerances and variability begins in the process design stage and continues throughout the life of the equipment. Maintenance operations must be coordinated with manufacturing to be integrated into the production schedules. As Itani also noted, there can be no effective production planning or customer promise dates with unreliable equipment.

---

*H. Kentaro Itani, "Why Is a Systematic Approach to Integrated Facility Management Crucial to World Class Manufacturing?" APICS National Conference, Orlando, Fla., April 1990.

Maintenance Imperatives

At the process level, there are four continuing pressures on manufacturing: developing total productive maintenance, achieving total quality, managing total cost, and accomplishing total time management—all to the firm's competitive advantage. IFM and its maintenance team must support them all.

Especially for high capital investment facilities, maintenance is of two sorts: routine (both planned and reactive) and turnaround (scheduled major outages). Routine maintenance takes place daily, off-shift or during scheduled nonproduction periods, and is carefully managed.

Example
ALCOA, Tennessee Operations. It is Wednesday. As scheduled, the hot line is down for weekly maintenance. Downstream, the buildup of hot-rolled coil is being withdrawn from the annealing furnaces while cold rolling and finishing are proceeding on schedule. Upstream, the casting and scalping of ingots is building up a buffer in the hot line soaking pits. On Thursday at 6 A.M. the hot line will crank up, and the entire plant will again be running toward synchronization.

In other parts of the plant, routine maintenance is taking place, either on the graveyard shift (for those activities not running) or on the sacrosanct twenty-first turn—midnight to 6 A.M. on Monday—before the regular day shift comes back on. How else can equipment that has already put in more than 20 years of hard work still produce to the rolling practice and specifications? The ALCOA hot line crew's preventive efforts are carefully orchestrated, and the timing is done in synch with production schedules to minimize capacity intrusions, as noted in Chapter 5.

ALCOA's annual turnaround planning is done much like a major construction project, with every activity choreographed and tracked on a PERT or CPM chart. The hot line is scheduled "down" for four to six weeks ahead of the "spring rush," with detailed project management techniques used to ensure an on-time finish. Other aluminum mills take on hot-rolling contracts (called *tolling*) and return the reroll coil stock to ALCOA for finishing. The Warrick facility in western Kentucky is a logical choice, and the two mills cooperate in scheduling their hot line downtimes.

Planning Requirements

Concept

Maintenance planning must fit the plant's philosophy of work: the kinds of support the equipment operators are capable of providing to the maintenance effort. The personnel plan for maintenance will obviously vary according to who does the routine cleaning, oiling, and adjusting of equipment. And the idea of preventive maintenance (as contrasted with reactive, or breakdown, maintenance) will also affect both crew size and work schedule. Ultimately, the goal is total productive maintenance (TPM), with the same sorts of benefits as those being achieved through total quality management (TQM).

Database

The supporting role for IFM (see Chapter 9), includes the maintenance history records, quality-related preventive or predictive maintenance records, and safety experience that bears on equipment—personnel interaction and related adjustments.

Demand Estimates

IFM must establish and develop an equipment history so that the plans can be compared with actual experiences. Equipment histories are also needed for parts procurement and, looking to the future, for evaluation of performance when equipment additions are planned. Predictive maintenance uses history or production wear rate estimates to perform maintenance economically at intervals, maintaining quality capability and minimizing the likelihood of a breakdown. The use of various nondestructive testing procedures and of monitoring devices (vibration and infrared heat detection, for examples) also can assist in predicting when equipment should be overhauled.

Statistics

A substantial amount of science is involved in maintenance planning: estimating the timing of various levels of lubrication, adjustment, recalibration, and parts replacement. Mean time between failure (MTBF) estimates are based on manufacturer data or on experience with similar equipment elsewhere in the company. These estimates are tied into the maintenance plan. Crew size and the level of workload regularly assigned will affect mean time to respond (MTTR) to unexpected requirements. The analogy with predicting service parts requirements is almost exact.

At ALCOA, such tasks as the regrinding and trueing of cold mill work rolls are always waiting. When mill maintenance crews are not engaged in preventive or reactive maintenance, they run the equipment that restores roughed-up work rolls to specification for reinstallation. If resurfacing is going on, it takes some minutes to drop everything and respond to a mill wreck or a work roll change.

Workload

Mean time for repair (MTFR) is also factored into schedules for maintenance and for manpower planning. Repair histories will improve MTBF, MTTR, and MTFR estimates and therefore the availability as well as the cost of maintenance.

Availability

Traditional maintenance practice is trading off the expected costs of maintenance against the costs of failure, with the goal of achieving a minimum expected balance between the two costs. This practice is being supplanted by the availability idea: the target now is to have at least 90 percent of all maintenance be preventive, scheduled maintenance. As implemented, this should result in more than 90 percent availability because not all random downtime will interfere with schedules.

Shared Procedures and Software

There are a number of similarities between production materials management and those practices applying to maintenance inventory, tools, and personnel.

Maintenance Materials Planning

APICS practitioners are familiar with material requirements planning (MRP), a computer-based system that can calculate net requirements and lead times for having materials available at the time scheduled for fabrication or assembly need. There are some direct analogies between the materials planning and inventory management approaches developed by APICS and those needed for proactive maintenance planning.

Consider that the MRP of a bill of materials (BOM) provides a detailed list of all the parts needed and their lead times to create the end item. If the item is a piece of equipment, identified by a serial number, and has a planned maintenance BOM (MBOM), then its presence in a maintenance master production schedule (MMPS) would be related to the parts required to do a particular level of maintenance and the skills and time to accomplish the tasks. The time required would be the offset, given the MMPS time bucket. The maintenance, repair, and operating (MRO) stores inventory file would reveal available parts, and the maintenance MRP (MMRP) could generate the net requirements and the lead time for procurement. And like the master schedule, MMPS won't execute without both parts and maintenance capacity, the latter described as the skills inventory records.

Planned Maintenance

The MBOM describes both materials and standard time required to carry out the planned procedures. When the work is scheduled, the maintenance parts and technician skills inventory files can both be examined for any net requirements that must be arranged before the procedure can be completed. The MRO stores files can be managed just like productive inventory. The technical skills inventory is, of course, based on personnel availability and any current or future allocations of those skills to other maintenance tasks.

Blanket parts purchasing, using a local source, can deal with both spot prices and long lead times. Putting equipment into the MMPS helps identify in advance possible manufacturing schedule conflicts, allowing for contingency planning. Useful for routine maintenance, this approach can also be combined with the MTBF and MTFR data in the equipment history file to determine what will be needed for major overhauls and thus protect the MMPS from overrunning. Where either time or work load favors subcontracting of services, the MBOM and the MMRP are useful in work specifications and time estimates. Preventive and predictive maintenance does, indeed, maximize availability.

Reactive Maintenance

The same MBOM can be used for emergency work, but it may require additional items. Breakdown can include equipment damage and resulting uncertainties. Here the notion of service parts inventory—statistically predicting parts consumption in the field as well as using MRP for those needed for assembly—applies to the breakdown MBOM and bill of labor. When a machine goes down, the expected parts and technical skills should be known and planned, even though the exact needs and timing are unknown. These critical parts are stocked or are immediately available by contract with an outside stocker/jobber.

JIT/QM and Maintenance

For the JIT/TQC operating environment, there is a customer-based mandate for strong maintenance, suggesting that the *direct* cost of failure is not the most important element to be considered. In fact, many feel that preventive maintenance costs should be treated as investments, not expenses.

In JIT systems, preventive maintenance has as its goal the minimization of unplanned downtime *failure* and the maximization of machine availability. Careful estimates are necessary: of practical capacity, the likely production plan in terms of machine hours versus production clock hours, and the times when production is not planned (between shifts, evenings, on weekends), when downtime will not affect availability.

Modern IFM requires sustained attention to the availability of equipment. This is not equivalent to uptime, although there is a belief in their similarity. "A bargain's not a bargain unless you need it" applies to equipment as well as sale merchandise. One needs to have equipment ready for operation when production work is scheduled and to carry out all maintenance when idle time is planned. This is a primary challenge for the maintenance team at any facility and requires significant coordination with manufacturing, quality, and production control personnel (e.g., the ALCOA Hot Line).

Computer-Based Maintenance Studies

The trend toward ever greater computer power and software, and the concurrent increase in computer familiarity, has meant that there are many supporting capabilities for effective maintenance management. Today, a variety of very useful techniques makes the fulfillment of Ken Itani's top priority task a reality.

As noted earlier, MRP is far more than a materials planning tool. It is, in fact, a simulator, with particular capabilities suited for maintenance management. Purchasing can use the same MMRP output to decide on sourcing for the next planning period and tell suppliers what will be needed and when. Maintenance management ensures that skills are allocated and manufacturing responds with downtime availability.

Quality

When quality and capability are added to the equation, equipment history files and MBOMs take on another significance. Experience, combined with the SPC data on the shop floor, leads to intelligent MMPS. The data base helps with the evaluation of move-out or move-in scheduling decisions based on current Production and Maintenance needs.

Total Cost and Time Competition

Cost considerations parallel those already discovered to be true for quality ("Quality is free"). Good planning prevents most surprises, and reduces the total cost of maintenance and of availability. Finally, good maintenance feeds productivity and minimizes unplanned outages. Time-based competition is enhanced.

Maintenance Support

Another important aspect is that of MRO (Maintenance, Repair, and Operating) stores. The "theory" has already been suggested. Here's an example:

In the Hanes Hosiery Mills in Winston-Salem, North Carolina, the repair parts for the knitting machines, located next to the machine room, were valued at over $2.5 million. But because of machine differences, the parts were not interchangeable, not even the needles.

Fixers to support the machines were on 24-hour duty. But with no capability to define a needed part and to retrieve it readily, the machines were not available! Frequently, this meant piece-at-a-time trips by the fixers, and premium prices for parts to repair idled equipment.

A significant effort was made, with the use of inventory control principles, to learn how to manage the large investment in parts. But only an incomplete history was available to help. This became recognized as a Hydra. Equipment choices needed standardization. Management came to see the need for preventive, as opposed to reactive, maintenance.

The expertise of the fixers gradually became a part of maintenance planning and was combined with the maintenance history on each kind of knitting equipment. In return for an exclusive contract, a local supplier agreed to stock both frequently and infrequently used parts, substantially increasing the equipment availability and reducing inventory investment.

Decentralized MRO Considerations

JIT suggests that tools, fixtures, and quality-check gauges should be located at the machines so that changeover items are reduced and the operators have better control. This can be applied to MRO as well. A decision depends on the plant's manufacturing concept. Issues such as decentralization, responsiveness, and personnel skill development, addressed in other texts in this CIRM series, are also relevant here.

Other factors must be considered. Are there physical security issues involved? Some plants use electronic test equipment, either in-line or at least available to operators. Is this equipment vulnerable to damage in the normal course of production operations? Is there a ready market for such equipment so that "shrinkage" precautions are necessary? How many equipment sets would be needed for each focus plant or process area? How will the test equipment itself be calibrated and maintained? What about infrared monitoring equipment and vibration-sensing devices?

How much "lore" is associated with the equipment and its repair parts? Does this mean dedicated technicians working alongside the operators to help tweak the equipment to maintain capability? Can the company afford a set of decentralized storerooms for MRO? Should storerooms be staffed? Will chaos prevail if they aren't? Where is the investment responsibility? How will replenishment be managed?

If foreign-built machines are used, as is frequently the situation, what are the lead times for parts? Are there any communications problems? Any import problems in obtaining parts in the United States? In overseas operations, repair parts for American-made equipment may be difficult to obtain quickly (politics, customs, and distance conspire to cause this). How can this best be managed?

TRENDS IN MAINTENANCE MANAGEMENT

The extrapolation of the various emphasis areas—preventive maintenance, computerized database use, scientific inventory management, availability and supplier partnerships—leads to some clear projections into the future for world-class companies.

Operator-Centered Maintenance

This trend makes sense from several standpoints. Apart from the desirable behavioral issues, no one can know more about potential difficulties than the person observing a process all day, every day. If operators can do their own machine maintenance, they feel that they "own" the machinery. Availability, quality, and total cost will be improved, and the repair time will be minimized. IFM will increasingly be involved in developing the operator skills toward this "transfer of ownership" and will be able, as it becomes a reality, to make staffing adjustments. TPM efforts will develop diagnostics through fault-tree analysis and improve the factory-recommended maintenance intervals through pre- and post-assessment of performance.

Design for Maintainability

IFM maintenance suggestions can help equipment manufacturers improve their product. Just as the product/process design team will strive for manufacturability, IFM will work with equipment and parts suppliers for repairability. Construction and sealing methods can facilitate disassembly and make reassembly mistake-proof, and the location of access plates can improve safety and make procedures easier to perform. Embedded sensors to measure critical temperatures or vibrations can provide early warnings of impending trouble. Material choices can extend maintenance intervals.

The familiar consumer analogy is that of doing routine automobile maintenance. If the product isn't designed for maintainability (in terms of timing, plug replacement, carburetor adjustment, radiator draining and refilling, access to the oil filter), endless hours of frustration can result.

TOTAL PRODUCTIVE MAINTENANCE

Another compelling trend in IFM is that of proactive, ever improving production systems maintenance, motivated by the positive results being achieved through total quality management (TQM) and by recognition that quality is only sustainable through maintenance. Total productive maintenance (TPM) is approached in a manner very similar to TQM's work teams, problem solving, workplace cleanliness and orderliness, and the generation and implementation of change-for-improvement ideas. From the lengthy earlier discussion on maintenance, IFM involvement is clear.

The TPM effort is supported by analysis of the equipment and of various failure modes and by experiments to determine actions that will prevent failure and predict when these actions should be taken. To measure MTBF, MTTR, and MTFR, the statistical approach is one aspect of the predictive-preventive approach to TPM. But it is not the only one.

Total Personnel Involvement

TPM was developed at Toyota, an originator of the JIT/TQM production system. The existence of strong, dedicated teams on the factory floor and in the offices made the move toward TPM logical and consistent with Toyota's other continuous improvement efforts. The efforts in production and in quality were based on cross-functional teams with a problem-solving approach to improvement. This parallels what is done to bring about TPM. *All* personnel eventually become part of TPM.

Six Big Sins

Seiichi Nakajima classifies the various elements that contribute to less than 100 percent good production as the Six Big Losses. Because the target is to be 100 percent effective in meeting production requirements, all six sins must be eliminated.

Equipment Failure

Through a combination of rebuilding, strong maintenance, and common sense, equipment should be restored to 100 percent (or better) of its original capability and reliability. Then predictive and preventive maintenance must keep it that way.

Setup and Adjustment

There are many ways to eliminate various elements of both setup and adjustment, and these are being practiced all around the world. The APICS literature has many good examples, some appearing in Burnham and Natarajan's *Manufacturing Processes* (1992). The ultimate approach is the one touch exchange of dies (OTED), demonstrated by Shingo (in Nakajima, 1988).

Idling and Minor Stoppages

Nakajima mentions sensor malfunction, delay of materials, and other nuisances that halt production briefly. Until the problems causing the delays are eliminated, the production schedule must allow for the stoppages or extend production hours to catch up.

Reduced Speed

When first seeking defect-free production, it is often a good approach to slow down and only work as fast as will allow zero defects (ZD), then gradually increase the pace as skill and equipment capability are improved. The goal, of course, is to operate in a ZD mode at design speed. (As synchronous-flow advocates have pointed out, however, it is silly to run at design speed if there is a bottleneck downstream and impossible to do so if there is one above.)

Defects in Process

Both rework and scrap reduce productive output. Random occurrence of defects is another Big Sin. TPM seeks to eliminate such events through problem solving.

Reduced Yields

Nakajima separates the randomness of failure of the production process and the start-up pieces that are run while moving to stable, on-specification production. This relates closely with Setup and Adjustment, above, because no adjustment should mean that the *first* piece of a new production part will be good.

The Logic of TPM

To reduce variability, to achieve previously impossible technical tasks, and, through automation, to combine processes and operations, production equipment has become highly sophisticated. But keeping this equipment *available* will still depend on human hands and minds. In a sense, TPM extends TQM, because maintenance wants no breakdowns and quality wants no defects.

Only by treating each problem separately can the total effectiveness of the process(es) be assured. Rather than keeping the studies statistical and reserved to the engineers, Nakajima urges that the entire team of operations, engineering, quality, and maintenance work together integratively and use Pareto approaches (focus on the significant few, rather than the trivial many) to choose projects for attack. The site team is the natural body to support such efforts.

How to Get There

For those companies embarking on the TPM path, the early stages are an investment in bringing equipment back to specification, then training people to use and maintain it properly. Beyond this lies increased profitability due to productivity, quality, and lead-time improvements.

How is this accomplished? In very much the same way as is TQM: through leadership, a strong fundamental understanding of TPM goals, involvement, small group improvement activities and suggestions, recognition, and dedicated effort. TPM includes maintainability improvement and expects much of the maintenance effort to be carried out through small groups of operators working together.

Proponents of TPM stress that, when put into companywide practice, it will minimize the life cycle cost of equipment by sustaining it at an optimal level. Like the good JIT production systems, the goal is the complete elimination of waste, including defects and downtime.

As with preventive medicine, TPM seeks from the outset to minimize deterioration, to measure it as it may occur, and to make repairs before failure happens. Cooperation between operations and maintenance specialists will accomplish this. Housekeeping and cleanliness are vital to TPM, because dirt is the enemy of all equipment.

A master plan for TPM promotion is shown in Figure 8–1. The figure is self-explanatory, but note that the three-year effort has six interdependent elements: to restore equipment capabilities, decentralize maintenance to operator teams, achieve quality in maintenance activities, bring equipment performance to the zero-defects level, develop ever-improving maintenance plans, and build operations and maintenance skills. There is substantial interplay among the various factors, and the sum is TPM.

Computer-Based Maintenance Systems

Earlier we presented some of the analogies between production materials management and MRO and other maintenance planning. There are a number of packages offering both models (for reliability, statistical studies, and replacement) and operations systems. Like SPC packages, the convenience and user friendliness must be traded off against the direct applicability of the models to the specific needs of the user. Companies generally have found that the packages (except for accounting control) do not mesh well with the needs for specific equipment-related preventive and predictive maintenance. Operator-centered maintenance and TPM, like operator-centered improvement activities, may be a more effective means of improving maintenance performance than computerized packages.

"Nonproductive" Maintenance

Personnel Areas

Office spaces, bathrooms, cafeterias, hallways, and storerooms also require maintenance. So do windows, parking areas, and landscaping of the plant site. Lamp replacement in both production and other spaces is part

FIGURE 8–1
A Master Plan for TPM Promotion

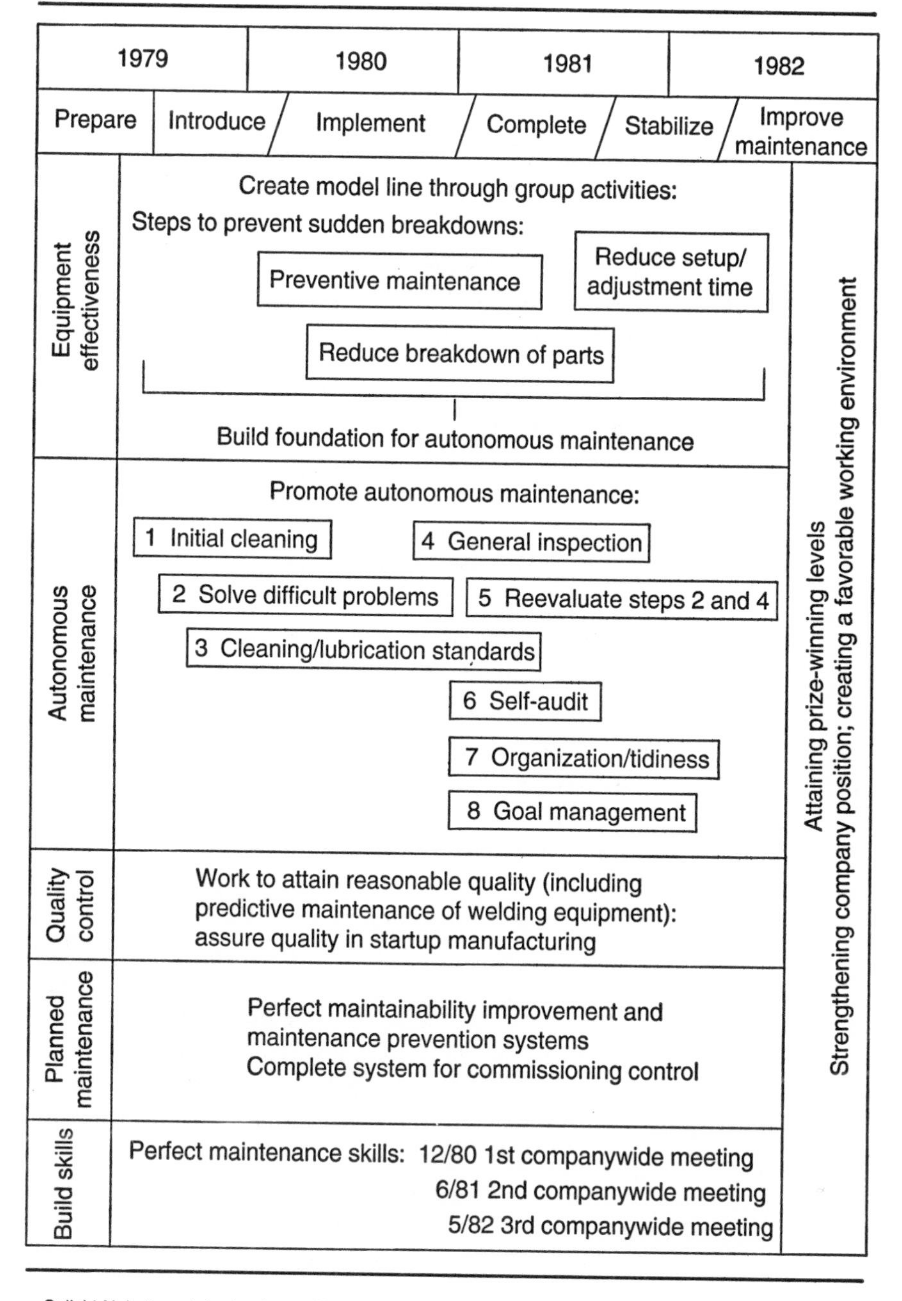

Seiichi Nakajima, *Introduction to TPM* (Portland, Ore.: Productivity Press, 1988).

of housekeeping. Whether carried out by subcontractors or by company staff, this kind of maintenance is as important as that of factory floors, equipment, and aisleways. The belief that cleanliness is everybody's job is reinforced—or destroyed—by how noncritical areas are maintained.

Electronics and Communications

In all parts of the modern facility, there are audio and video services providing background sound, news and performance data that help support the climate for improvement in the plant. IFM is generally responsible for this, either through an outside contractor or the maintenance team, and like the other public spaces maintenance, these audio and video services can set the tone for plant quality.

LESSONS FOR THE INTEGRATIVE RESOURCES MANAGER

The performance of the industrial facility depends on its resources: capital, materials, management, manpower, equipment, and information. Failure in any area leads to performance degradation. Excellence *is* as excellence *does*.

Experience suggests that the most ignored resource today is that of the skilled and motivated maintenance team, perhaps taking the place of the "whipping boy" quality of earlier decades. Sadly, trade-offs between higher-quality products and lower-cost production are still sometimes presented as necessary. Similarly, maintenance is presented as the cause of production delays and overtime costs. And just as surely as companies are confronted with the costs of poor quality, they are also faced with the costs of inadequate maintenance—whether recognized or not.

Effective maintenance is critical to operator safety and morale, production schedule fulfillment, and product quality. Maintenance is therefore central to any plan to improve performance and gain market share. It requires much more planning and execution management than the production process itself because there are many more dimensions: equipment, parts, personnel numbers and skills, quality characteristics and required capability, and systematic interdependencies among sequential steps of the production process, among others.

World-class status and its mandated total quality emphasis have both immediate and long-term implications for maintenance management. In the short run, quality needs will lead to mandatory proactive maintenance programs driven by process capability needs. In the long run, maintenance, like quality, will be decentralized to the machines and their operators, but with strong support from the related parts of the organization. This support must include human resources, finance/accounting, information systems, production, scheduling, materials management (both inventory and purchasing), product and process development engineering, and the site development team (for both quality and maintenance). Continuous improvement will span both TQM and TPM, and all will target the internal and external customers for the products being created.

REFERENCES

Bell, Robert R.; and John M. Burnham. *Managing Productivity and Change*. Cincinnati: South-Western Publishing Company, 1991.

Burnham, John M.; and Ramachandran (Nat) Natarajan. *Manufacturing Processes Student Guide*. Falls Church, Va.: APICS, 1992.

Hayes, William H.; and Steven C. Wheelwright. *Restoring our Competitive Edge: Competing Through Manufacturing*. New York: John Wiley and Sons, 1984.

Itani, H. Kentaro. "Why Is a Systematic Approach to Integrated Facility Management Crucial to World Class Manufacturing?" Presentation. APICS National Conference, Orlando, Fla., April 1990.

James, Robert W.; and Paul A. Alcorn. *A Guide to Facilities Planning*. Englewood Cliffs, N.J.: Prentice-Hall, 1991.

Nakajima, Seiichi. *Introduction to TPM: Total Productive Maintenance*. Portland, Ore.: Productivity Press. English translation, 1988.

Tompkins, James A.; and John A. White. *Facilities Planning*. New York: John Wiley and Sons, 1984.

CHAPTER 9

IFM OPERATIONS AND TRENDS

Once the plant goes into operation, a whole new set of challenges arises. Insurance, special safeguards, disaster planning, ongoing support for the many improvement activities, carrying out the maintenance plans, housekeeping, recruiting and training, keeping data files accurate and current, and a myriad of other tasks occupy the site team in its new role.

IFM AND THE FUTURE

Since IFM plans, installs, and maintains the site, facility, and equipment environments necessary to support the firm's long-term business objectives, it has a *representational* role in corporate planning, an important *enabling* role in developing and then implementing the facility plan, and a vital operations *supporting* role, while tracking trends and preparing for events which will affect the plant in the future.

The focus for this chapter is suggested by Figure 9–1.

Explicit enumeration of *what, why* and *how* is critical, requiring consideration of all relevant internal and external factors, including location and design. Figure 9–2 summarizes these external and internal factors and suggests their interactions.

OPERATIONS

Both greenfield and grayfield construction programs can be handled by traditional project management methods as detailed in Chapter 7. As with all good management, the goal is to finish on time, within budget, and certainly with a completed facility that *meets the customers' needs*.

FIGURE 9–1
Roles for Industrial Facilities Management

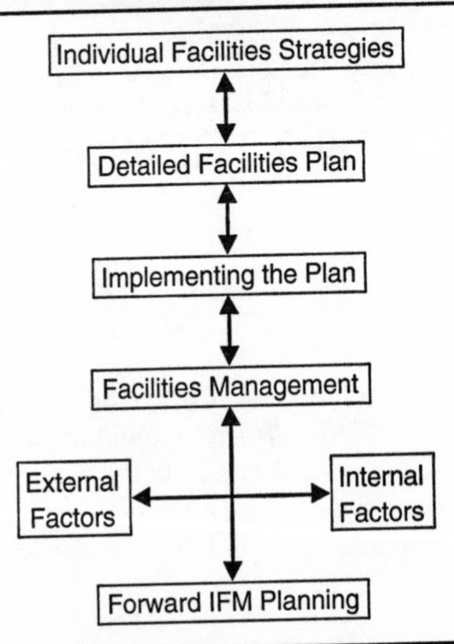

Source: John M. Burnham and Ramachandran (Nat) Natarajan, *Manufacturing Processes*, Student Guide (Falls Church, Va.: APICS, 1992), Fig. 7–1, p. 7–2. Used with the permission of the publisher.

The move from planning to operations is a gradual one, beginning with the planning associated with maintenance—an activity tied into production. Database requirements to allow for engineering (equipment and site/facility plans) are also developed before operations commence. Planning and operational concerns for IFM and the site team include emergency planning, regulatory compliance, safety records, waste and by-product management, site housekeeping, traffic routing, and industrial and personal vehicle parking. One successful team effort is illustrated in the following example, summarizing much of what the previous chapters have presented.

A Facilities Development Example

In 1963, when the Farmville, North Carolina, plant of Collins and Aikman was first constructed, its product assignments were principally commodity knit-fabrics, items that would sell at retail for under $2 a yard. There were a relatively few standard widths, and fabric design was simple. The process was straightforward: Creels of yarn were received in

FIGURE 9–2
IFM Interactions

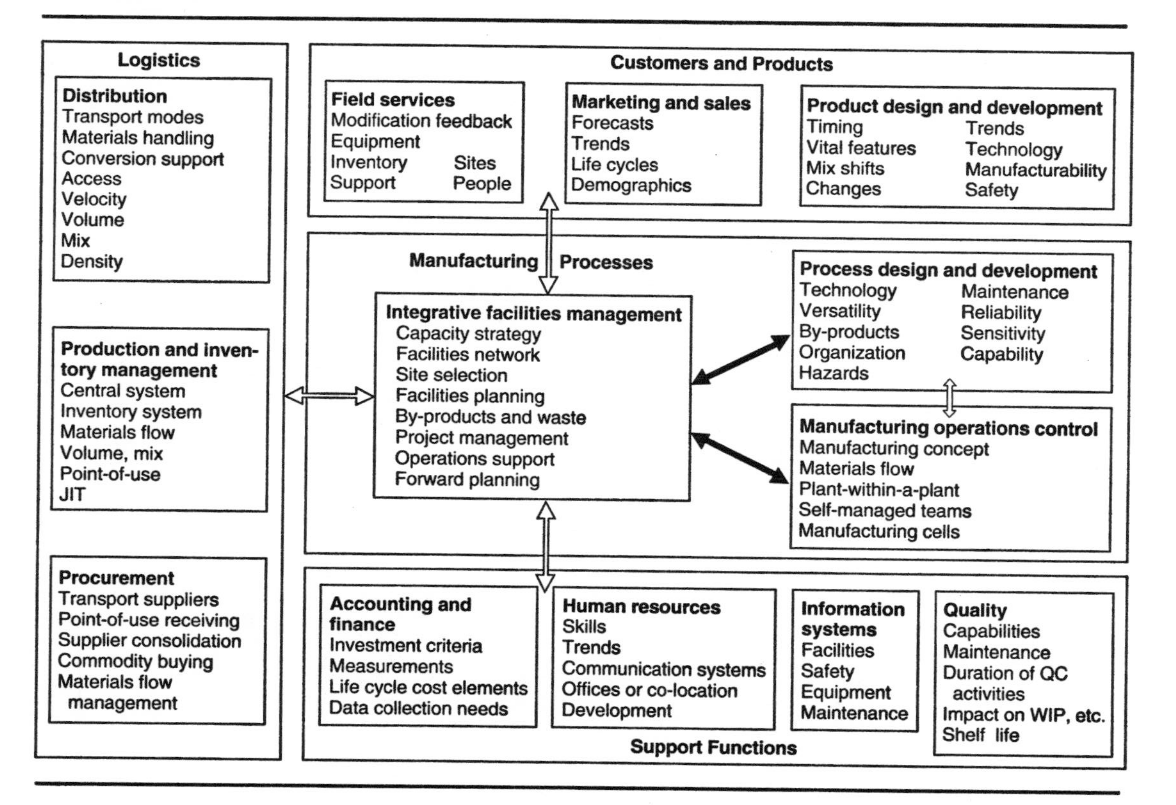

large boxes. The right number of ends were led to a warping beam, where yarn was wound in preparation for knitting. The beams were then doffed (manhandled) onto the knitting equipment, already set up for the fabric style. The ends were led across the bars so they could interact with the needles, and the machines operated continuously until the yarn ran out. Another beam was doffed, and the process began again. The undyed knit materials (griege) were stored until there was demand or until a dye run was scheduled. The fabric then was wound onto perforated beams for immersion and a dye bath forced through the center of the beam until saturation was complete. Drying, cutting to processing length, and then stretching and heat setting on tenter frames completed the process. After inspection, the bolts were sent to the customers.

According to Gene Crawford, the original site manager, considerable time was invested in future expectations when the facility was being designed. A team did contingency planning, thinking through all the possible combinations of products, mix, and volumes that might eventually be sent to Farmville. The greatest challenges related to utilities: electricity, gas, water, and drains. These would be expensive and, once installed, almost impossible to modify. Significant power for lighting and equipment would be required for warping and knitting, especially for machine setup. Dyeing and finishing operations would require large amounts of water and gas or electricity, as well as drains for the expanded beam dye baths. It was decided to cover these expectations with oversize water, drain, and gas lines, even at greater initial cost.

As a result, more than 20 years later the C&A Specialty Fabrics Division at Farmville was generating over $100 million annually, with a return on assets of over 20 percent—quite astounding for the rapidly declining U.S. textile business. The product mix had shifted from commodity to high value added fabrics: lingerie, actionwear, and automotive seat and headliner material. Nearly 200 high-speed knitting machines were going full blast, as compared with the original 60. Some wire-guided automated materials handling vehicles had been added, and a number of plant additions had taken place, but the core facility was still basically intact. Steam, gas, and water lines were for the most part undisturbed, except for extensions to service new equipment. This was true even though dyeing now was done by giant washing machines using venturi jet action, able to process 500–1,500 yards of heavy fabric at a time. A drawing of this plant in 1986 is shown in Figure 9–3.

When a JIT/TQM effort was begun in late 1986, only procedural changes were required to bring a two-week lead time down to less than

FIGURE 9–3
Plant Layout of Collins and Aikman Farmville Plant

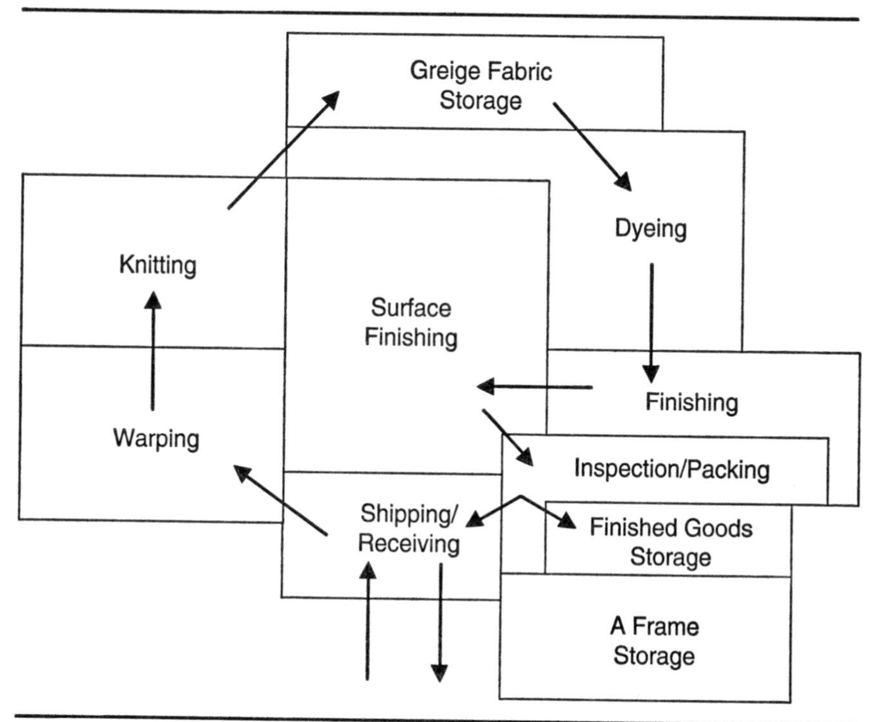

Source: Robert R. Bell and John M. Burnham, *Managing Productivity and Change* (Cincinnati: South-Western Publishing, 1991); figure adapted for John M. Burnham and Ramachandran (Nat) Natarajan, *Manufacturing Processes,* Instructor Guide (Falls Church, Va.: APICS, 1992), Fig. 7–6a, p. 7–15. Used with the permission of the publisher.

four days from griege goods to pack-and-ship. Quality awareness efforts began to bear fruit. Gain-sharing on productivity and quality improvement was established. With improved practical capacity, the increase in orders brought the return on assets to over 30 percent, with no capital equipment additions. The core facility *still* remains intact.

The Plant

Central to the existence of the IFM function is the plant itself and the kinds of future services, support, and planning inherent in IFM. These are, for the most part, "technical" considerations, but they are critical to plant competitiveness (see Figure 9–4).

FIGURE 9–4
Roles of IFM for an Existing Plant

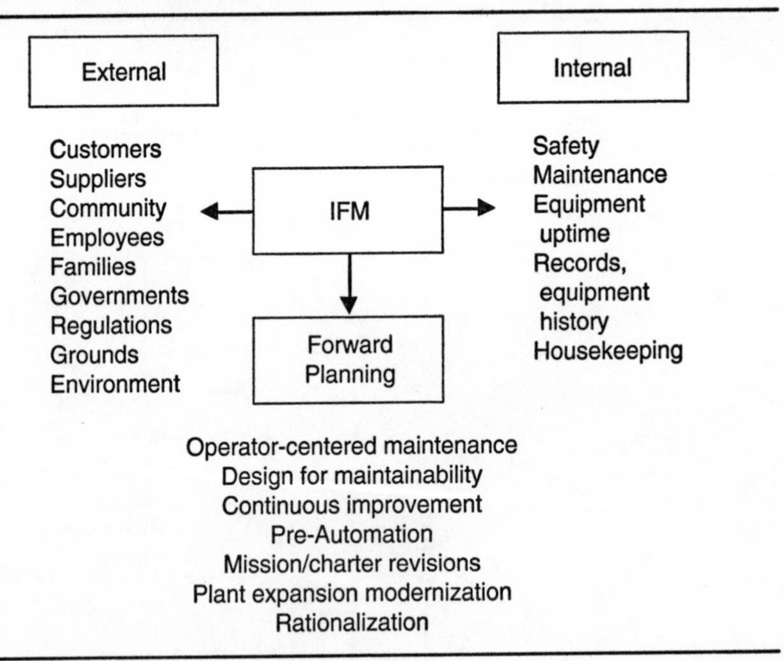

Source: John M. Burnham and Ramachandran (Nat) Natarajan, *Manufacturing Processes,* Student Guide (Falls Church, Va.: APICS, 1992), Fig. 7–10, p. 7–28. Used with the permission of the publisher.

Computer-Based Maintenance Systems

Chapter 8 introduced some of the analogies between production materials management and MRO and other maintenance planning. Like SPC packages, the convenience and user friendliness must be traded off against the specific needs of the user. Companies generally have found that software packages (except for accounting control) did not mesh well with the needs for specific equipment-related preventive and predictive maintenance. Operator-centered maintenance may be a more effective means of improving maintenance performance than computerized packages.

Total Productive Maintenance

Proactive, ever improving production system maintenance, paralleling total quality management in emphasis and methodology, is increasingly being implemented. This effort is supported by analysis of the equipment and various failure modes and experiments to determine actions that will prevent failures and predict when these actions should be taken to avoid them. Like good JIT production systems, the goal is the complete elimination of waste—in this case, unscheduled downtime and incapable equipment.

Computer-Assisted Facilities Management

Logically, everything that is being accomplished in computer-supported engineering product and process design and in the direct digital toolpath programs (call this *manufacturing CAD/CAM*) should also be available to help keep the site development and plant engineering records current and accurate. There has been considerable progress in doing exactly this for business office design and maintenance (call this *office CAFM*). There are some APICS reports stating that the newest CAD/CAM systems have office furniture icons and specific equipment "footprint" sketches, allowing the site team to keep the plant layout up to date. This author's research shows little current crossover between the two, despite the obvious advantages. As the concept of a unified data base translates into operational terms, this will undoubtedly be of increasing importance.

IFM Databases

IFM clearly must develop and sustain excellent records. Principal among them is the ongoing plant mission and charter, which drives the planning effort. Actual as-built facilities drawings will support any and all maintenance or modification work required. Current technology allows electronic storage and retrieval (imaging) of facilities plans and quick, accurate updating, using CAD. Tying the IFM CAD needs into the product CAD needs offers great potential savings.

The equipment records and maintenance history serve as the basis for MMPS, MBOM, and MMRP procedures. These records will also help in other ways: equipment and personnel safety experiences, evaluation for additional equipment purchases, and sustainable capability studies. The TPM thrust and its preventive maintenance antecedents both require the use of monitoring equipment for atmosphere, noise, and hazards. These also need electronic data files to facilitate investigation of failures, causal analysis, and consequences.

Personnel and Plant Safety and Security

The IFM database must serve as protection and as help with prevention. Detailed understanding of processes and the material and by-product elements will permit both demonstrated compliance, through material safety data sheets, and the development and posting of hazard warnings to protect the contact personnel.

The logical extension of safety, hazard warnings, and other personnel protections is that of fire-fighting equipment, sprinklers, and evacuation routes. The locale (is it hurricane, tornado, or earthquake prone?) or the nature of the product (are there explosive or highly flammable chemicals?) will lead to emergency planning and exercises to maintain preparedness. As a part of the fixed plant and equipment, there must be appropriate alarm terminals, manual fire-fighting equipment, and, in most cases, automatic sprinkler systems. The insurance classification of a facility can be greatly improved and its premiums reduced through the proper combination of construction, ventilation, and fire-fighting equipment. With office personnel co-located to the production floor and a trend toward training, health, and even child care facilities at the site, the need for rigorous safety and emergency planning is great.

Legal liability or regulatory challenges to safety or environmental compliance will be directed to IFM, and good records and activity logs will be vital. The formal safety program and associated records may also fall to IFM. Most plant visitors notice the Green Cross of Safety displays noting the number of days since the most recent lost-time accident.

Continuous Improvement

World-class manufacturing is driven by competition and customer expectations. Within the plant, the combination of leadership, challenge, trust, involvement, learning, experimentation, change, and making the changes work creates an exciting and dynamic environment. Substantially all the changes will affect the IFM function. The flexible treatment of utilities discussed in Chapter 5 will bring power, ventilation, other services, and materials handling to a reconfigured layout with minimum disruption and minimum cost.

The existence of any production operation involves these *supporting* activities and changing needs. When one adds continuous improvement, the amount and variety of support increases and stays at a high level. The many changes that accompany worker-centered management, problem solving, quality, and maintenance, lead to multiskilled, multimachine operator capabilities and multilevel task assignments. Both physical and psychological support should be present—meeting spaces, rest areas, and secure storage on the one hand, and pleasing surroundings, quality lighting and ventilation, and excellent noise abatement on the other—to provide the background support that will stimulate even further progress.

At the same time that more space and better appointments will increase both cost and facility size, there are favorable space impacts from JIT and TQM. The overall level of inventory will be reduced and the space freed up for other purposes. And because of the improvements in quality, rework may be eliminated as a plant routine. Thus "cripple" stalls or aisleways can be reassigned. On the other hand, decentralization and self-management generally mean more test equipment, gauges, and tools located near points of use; and for multiple equipment locations, this will mean redundancy, with both more space and more investment required.

Technology Dynamics

Many plant changes arise from bottom-up improvements. Others, however, are driven by the opportunities to improve productivity and quality, through technology enhancements. These might include more flexible or more capable equipment additions or exchanges, or the implementation of new procedures or software, to provide improved control or communications.

The use of programmable controllers for managing machine tools is growing rapidly. These require different personnel capabilities (electronics versus electromechanics) and must be planned for. Expert systems are increasingly being used to help with teaching-learning processes and to facilitate fault diagnosis. For example, Abbott Diagnostics Division provides 24-hour support to customers for their blood analysis equipment, and skilled medical technicians use diagnostics to help get the user back "on line" without the wait for a site technician visit. Abbott is moving rapidly toward expert-systems support.

Pre-Automation

With a focus on process perfection, world-class manufacturing is moving toward the elimination of most cost-added (waste) activities. Inherent in the disciplined progress toward the flexible, mistake-proof, waste-free plant are simplification, flowlike manufacturing, the elimination of bottlenecks, and teamwork toward continuous improvement. First, the manufacturing system must be cleaned up, both figuratively and literally. Then a process may become a candidate for automation. Work station reorganization for tools, fixtures, and visibility involve IFM support, as do all the stages of pre-automation for both procedural and physical changes.

Mission and Charter Revisions

Considerable change will take place over the life of the facility. Some of it will be local, in reaction to regulatory mandates or community circumstances, but more will be in response to sustained improvement pressures and to shifts in the role of the plant in the total manufacturing system. These shifts come mostly from external factors—competition, customer expectations, technology changes—and they are reflected in revisions to the plant mission. IFM must track the environment and coordinate closely with marketing and product designers, for changes here can affect the focus of the plant.

Plant Expansion or Modernization

To maintain their advantage, successful products and successful plants will go through a number of changes, adding to capacity and improving manufacturing technology. These changes should be made congruent with the capacity and facilities plans supporting overall corporate strategies.

Rationalization and Downsizing

When products are in the mature or declining phase of their life cycles, there will be great pressures on manufacturing, especially for cost reduction. Cutting the direct labor force is the traditional response, "to bring the head count back into line with volume." A lower aggregate labor cost does improve the balance sheet, but cutting the work-force stymies efforts at human resources development, involvement, and continuous improvement. Nonetheless, for the product lines involved, there seems no practical alternative but to downsize. Again, IFM must assist in facility reorganization to match the new scale of operations and staffing level.

Rebirth

When old products decline, the alternative to downsizing is to add other, new products. These sometimes are the by-product of plant employee ideas or of product development teams seeking ways to help the facility survive. As with all other plant change programs, IFM must both *enable* and *support* creative uses for existing equipment and process configurations, as well as work with manufacturing to address the retraining effort for both operations and maintenance personnel.

Warehousing and Transportation

There may be many more parts and product storage facilities than there are manufacturing sites as a practical matter (though this is changing with focus, smaller plants, and just-in-time). The existence of a distribution center is not the equivalent of a smoothly operating one. Inappropriate distribution facilities mean poor customer service or very high inventory and operating costs—or both.

Good systems analysis will help, through examination of the flows required by both internal and external customers. Materials handling studies will help determine basic commonalities: the unit load, the basic production quantity, the basic out-of-plant movement quantity, the current and expected rate of consumption by product/item/group, and will be used to optimize the associated systems.

Warehouse Organization

Operationally, the choice and number of storage locations depends on the criteria for performance. A recent visit to a major health care company's distribution center complex showed *fixed* storage regions by major customer, some *random* locations for mixed items, and a Kanban *pull* system for transport among warehouses to stage full outbound truckloads. Unlabeled "bright stock" made up the buffer to meet unexpected customer needs. Continuous improvement through teams was also much in evidence.

Manufacturing Warehouse?

Warehouses for JIT manufacturing sometimes incorporate preliminary manufacturing operations-blanking or cut-to-length operations. In distribution, it is not at all unusual to have a break-bulk operation taking place at the distribution center or warehouse. The health care manufacturer used its DCs to convert bright stock into customer-specific labeled products. As with the manufacturing facility, warehouses must match the purposes for which they are built, and be kept current throughout their lives.

Transportation

Carriers are now focusing on being world-class providers of many services involving consolidation, shipment tracking, break-bulk, and point-of-use delivery. Each of these entails expert analysis of the collection and delivery needs to define what company facilities will be required to manage the assignments.

In the JIT transportation business, Roadway Express uses its terminal facilities to regroup large mixed orders from a single supplier to match a single customer's multiplant needs—this for many shipments a day and for several classes of customers.

Federal Express manages its hubs for overnight delivery of each *individually tracked* customer lot, with quite literally millions of pieces of urgent information and small packages. United Parcel Service does this for both land and air shipments of intermediate size and weight.

Some manufacturers have integrated their operations with the distribution centers of their *customers* and manage both manufacturing and distribution replenishment. "Quick Response" in the textiles and apparel markets is but one example. This requires an equivalent understanding of how IFM must support both the company and customer facilities in terms of the movement of goods.

IFM TRENDS

Change is everywhere- in hardware, software, and "brainware." Maintainability is becoming a design feature. Lowest life cycle costing has led to more outsourcing. Faster, less expensive computers have helped with database accuracy and reduced maintenance costs. Both process and product liability issues place great pressure on IFM to assure proactive compliance with regulation and to be able to prove it. Integration through the database may facilitate integrative thinking among physically separated groups in the plant.

Change

There will continue to be many pressures on manufacturing: maintenance, quality, cost, and time, all of which mandate continuous improvements. Product and process redesign and manufacturing system revision require IFM involvement and proactive planning. In addition to the continuous improvement of product and process and manufacturing infrastructure, IFM must accept primary responsibility for other internal and external factors (not nearly so glamorous as product-related ones) that can also lead to plant shutdown or even termination of operations.

Design for Maintainability

IFM maintenance suggestions can help equipment manufacturers improve their product. Just as the product/process design team will strive for manufacturability, IFM will work with equipment and parts suppliers for repairability.

Outsourcing

The use of lowest life cycle cost has added to the cost pressures on IFM. Many specialist firms have developed methods that challenge the tradition of having company employees carry out janitorial, custodial, grounds-keeping, and security services. Further, MRO purchasing adds to workload and is not nearly as challenging as buying for production needs.

"The outside shop" has always been available as a means of obtaining the skills of another company supplier. Make-or-buy is a usual decision process for parts purchase. Major *services* outsourcing trends will affect IFM's future scope. One trend is to have IFM, or the equivalent site development or plant engineering group, manage the subcontracted services. This requires development of technical and operational specifications, bid evaluations, and recommendations for action. It also means administering the contracts, once established, and providing performance feedback to the contracting firm. User inputs, modification of contract terms, and decisions on contract continuation will generally fall to some subset of the IFM staff.

Security Services

Having company employees provide physical and property protection at the site is being rapidly displaced by Wackenhut, Brinks, Rollins, and other companies, which companies can *design* the security arrangements as consultants to IFM, then recruit, train, equip, and manage security employees, providing bonded services on a contract basis.

Mail

Postal sorting and delivery services and bulk-mailing processing are other areas where outside services, rather than internal ones, have been shown to be more economical.

Consumables

Rack jobbers—used for decades in supermarkets to replenish specialty snacks and hang-on rack items—are now performing the same services for a wide range of MRO items. Gloves, pipe fittings, nuts and bolts, gasketing blanks, sealants, and so forth are routinely resupplied and invoiced without actual ordering processes. The benefits include cost, availability, and the ability to have store supervisors focus on more critical or costly items. The *customer* sets the bogey (restocking levels), and the jobber complies. Unfavorable experience leads to exception discussions to improve the bogeys.

Technical Specialists

Shared specialists—those technical equipment and services suppliers who maintain the expertise and the parts to carry out services for sophisticated equipment—can and do bid against an inside estimate for this work. For example, a knitting-machinery distributor might agree to carry stock items to be used by a number of different textile manufacturers in an industrial park, making it unnecessary for the individual textile mills to keep anything but unique item stocks. This was true at Hanes and at Collins and Aikman, both being located where several other substantial textile firms were also contributing to the repair parts demand.

Computer Facilities Management

This is a primary target for outsourcing in 1994. Outside EDP facilities managers have the ability to add or reduce staff to "chase" more closely the computer operations and maintenance workload. They can also handle the equipment upgrades, additions, and downsizing activities that characterize this rapidly changing service. If the contracting firm has the appropriate know-how, there are both economic and quality advantages.

The site team almost certainly will have outsourcing decisions to deal with, both in planning and in operations support. IFM will likely have the management task as well.

External By-products

The initial facility plan considers many of the flows across plant boundaries: people, supplier materials and service personnel, distribution transport, water, drain and sewage lines, and waste removal. These are treated through systems analysis, and provisions are made to provide support (see Chapter 5).

But these flows are dynamic, as are the support services required through IFM. Changes in the number of office or factory employees leads to space adjustments and personnel relocation. Changes in duties—cross-training, operator changeovers and maintenance, and support staff on the factory floor—require layout modifications and perhaps workplace redesign for ease of access and multitask or multimachine operations. Altering the logistics system (to JIT, for example) can lead to both internal and external storage and material routing changes. Equipment usage can change as the product mix varies, and associated material volumes may also be altered. Changes in the cost or availability of utility services or in regulations affecting the flows on the site or the wastes from the site call for monitoring and effective response in support of the changes.

Waste

Disposal sites have become very unpopular. Today, industrial tax incentives may lead to considerable ingenuity on the parts of IFM teams to find ways to sell all or more of their waste volume to others who can use it. And the trend will continue. Steel and aluminum containers have become "collectible." And chemical wastes, formerly dumped into sewers or gravity drains, are now heavily taxed or regulated, motivating internal recycling and reuse.

Recycling

A hot button for industrial and residential citizens, recycling must involve IFM during design and through operations support. Paper plants have shifted to treating and reusing the wood chip chemical "liquor" and to the purchase of recyclable "waste paper." Textile plants precipitate solid wastes and dye residues, thus recycling cooling and wash water and greatly reducing wastewater volume. Collapsible, returnable containers are a growing trend.

Pollution

Airborne waste is an active EPA target, and acid rain a matter of international concern. Is the solution the retrofitting of millions of automobiles with catalytic converters, the redesign of the power source (like Honda's CVCC engine), or the use of another fuel (battery, natural gas, or steam)? Natural gas already is used as boiler fuel to "purify" power plant stack gases, bringing the emissions down to regulatory standards.

Postproduction Liability

A significant shift in public opinion and in government regulation has meant continuing manufacturer responsibility for the product (or the by-products of that product) after use by the consumer. Recycling of lead-acid storage batteries, automobile tires, and metal and glass containers is already common. It's likely that many other products will be added to the list, for one reason because solid-waste disposal sites are under mandate to reduce landfill by 25 percent by 1995.

Insurance

More than facility insurance is involved when by-product volumes, vapor emissions, and health issues are present. The insurance coverages may be a matter of company policy and merely carried out by IFM. But projections in terms of contingency planning may rest more with IFM than with any other plant-based group. All of the foregoing concerns carry a financial risk and an opportunity to seek protection.

Secondary Products

There is an increasing sensitivity to environmental issues, and one interesting derivative of that is by-product development and marketing. Dual pressures make by-product development certain: the cost and disposal problems associated with waste and the desire to eliminate environmental degradation, acid rain, and global warming. The challenge is to use the expertise already developed by the site development team to run the process backward—that is, look at ways in which process changes can alter the form, composition, or amount of offal and find another customer who actively consumes something like the offal being produced. This challenge needs to be met without hurting the primary product's attractiveness to its consumers, and it must involve marketing and other functional specialists.

In all these cases, the long-term solution has to lie in systematic rethinking. The existence of Nucor and other mini-mills as successful producers of increasingly high-valued products is testimony to the creative use of junk automobiles: too low in value to transport great distances, a huge blight on the landscape, and apparently, an assured supply source so long as high accident rates and obsolescence continue.

Human Resources Management Issues Affecting IFM

Workplace literacy is perhaps the largest single challenge to meeting 21st-century demands, with poor health practices and substance abuse close behind. Many of the enlightened practices coming into maturity today depend implicitly on considerably greater formal learning than has ever been the case before.

Exxon Refining uses worldwide satellite tracking to provide information on performance at any Exxon refinery. This means that day-shift technical operators can "attend" the lights-out (unattended) operations in other time zones, because the same performance data apply to each plant.

The automated Allen Bradley plant in Milwaukee, Wisconsin, is completely managed by programmable controllers. Rather than toolboxes, the "maintenance staff" carry control keys and computer diskettes with diagnostics recorded on them!

Maintenance

Technological trends cause some critical issues in learning. In traditional machining businesses, newer machine tools are almost entirely CNC/DNC equipment. Frequently, machine complexes operate in coordination as flexible manufacturing systems. So the traditional mechanical, electrical, and fluids technicians have become cross-trained among all these crafts. A number of technical institutes now provide graduates with skills in both programming and maintaining robots.

Learning

Three specific aspects are of particular concern to proactive IFM: minimum skill levels (literacy and numerical measuring), changing skill requirements (equipment and tasks), and the learning curve and continuous improvement. The challenge is similar in all three cases: to ensure that the needed skills are defined and that programs for developing them are made available. This applies to the operators who will increasingly be doing routine equipment maintenance and to the skilled plant maintenance force.

Cures?

The Hach Company, a Loveland, Colorado, maker of water analysis equipment was finding that 70 percent of job applicants were being rejected because they couldn't write a complete sentence. With the shrinking pool of entry-level workers, Hach began offering its 820 employees courses in basic writing and math.

Hach now invests 9 percent of its $21 million payroll in training. It offers 42 courses, has three full-time teachers, and taught 10,000 hours in 1990—all on company time. Work error rates have dropped, and Hach posted record sales and profits each of the past seven years. Turnover is down from 11 percent to under 7 percent, and as President Kitty Hach puts it, "This is the future. If people are educated, it makes running a business that much easier."

At a recent workshop, a human resources professional asked about hiring criteria for today's manufacturing and what skills should be sought. After a good bit of discussion by the workshop participants, they concluded that hiring should target the best people in terms of their potential for absorbing training on a continuous basis, not just those having a specific experience base.

LESSONS FOR THE INTEGRATED RESOURCES MANAGER

IFM must provide effective *support* to changes resulting from both external and internal forces. The presence during the planning phases of many nonproduction, nonengineering people on the SD team provides for a combination of networking, monitoring, and good breadth of skills to carry out these changes during the plant's operating life.

One unique aspect of the SD or facilities management team is that the team develops the *facilities product*. Plant performance is measured not only in units produced but in *how* they are produced—cost, quality, volume. Another future performance measure is likely to be the number of engineering changes required to fix a plant design or layout glitch. Plant design flexibility and modularity will be tested for effectiveness throughout the life of the facility.

Today's world is full of shifts in regulatory stance—almost never toward relaxation of current standards. At the same time, court decisions and statutes combine to burden the manufacturer with increasing responsibility for both product and process performance—even after the product's useful life has ended! IFM must work with product and process designers to create value rather than waste from by-products. *Green engineering* is one of the terms being applied to this practice, and it seems sure to increase in importance in the coming decades.

Competitors bring significant pressure to bear on the plant, with shorter life cycles causing both process and product redesign. Parts of the manufacturing system are faced with almost daily change—through continuous improvement, the product life cycle, and needs-driven modifications.

Simpler, more effective planning and control systems are being implemented in world-class factories. Logistics is becoming the companywide equivalent of the materials management function at the factory itself. There are profound facilities implications for all of these changes, especially in layout and materials flow.

Perhaps the strongest single support emphasis is the role of maintenance in maximizing the availability of capable and reliable production capacity. Without effective maintenance, neither quality nor quantity can be ensured, especially on the customer's schedule. Total productive maintenance development has the advantage of further involving the whole work force in improvement and self-direction, these already being strong trends because of JIT/TQM.

Awareness of the support requirements that continuous improvement brings is particularly important in view of the increasing breadth of responsibility assigned to IFM. Modularity and flexibility in design mean lower total costs in expansion and modification. More actual redesign by the work teams on the factory floor benefits all concerned.

The education and skills needed to operate a modern factory effectively have placed much of the training and upgrading responsibility on the company, rather than on formal academic systems. Operators must learn how to maintain their own equipment. Maintenance technicians must not only learn about new, electronically controlled equipment in order to maintain it, but also how to help transfer these skills to the operators. And management must deal with the 85 percent of the improvement opportunities *not* under the control of the workforce!

Checklist for IFM Operations Support

Facilities-related results depend on many features already mentioned. A checklist of important features would include the following:

Point-of-use storage for

- Production materials
- Tooling and fixturing
- Changeover tools and devices
- Gauges and other quality measurement devices
- Test instruments

Point-of-use locations for

- Bulletin boards
- Problem boards

- Individual-skills progress boards
- Training areas

Visible controls for

- Line stops
- Material shortages
- Machine malfunctions
- Production and quality updates

Communications facilities for

- "News"
- Performance
- Stock market
- Public interest broadcasts
- Presidential and management meetings
- Team projects and accomplishments
- Progress reports and departmental results

Finally, there is the general mandate of constant interaction with external and internal suppliers and customers to meet their changing needs in a cost-effective manner.

APPENDIX

A FACILITY TRANSFORMATION EXAMPLE JCI/PIKEVILLE PLANT

In 1983, the Ferro Manufacturing plant in Pikeville, Tennessee, was the most costly of three company locations, all of which supplied the automotive OEM market. Operating traditionally, there was considerable personnel turnover, lead times were long, and quality was mediocre.

A new plant manager, transferred from the Lexington Ferro plant, began a transformation by bringing in Ken Wantuck, a Bendix materials manager-turned-consultant, for a brief visit. Ken affirmed that a process of change would be both possible and profitable. Over time, the plant moved from assembly lines to product-related manufacturing and assembly cells. The plant's engineers worked with the departmental operators to lay out the new work areas and to modify them as needed. Supervisors worked together on a monthly basis to determine the number of people needed in each department to meet production requirements. Supervisors and departmental operators did the rest. Over 70 percent of the operators could not only do every job in their own area but across the whole plant!

Housekeeping and lighting were given high priority, and visible controls became the rule. Point-of-use storage was established right at the workstations, and forklift trucks and big shop carts all but vanished, replaced by small, uniform containers. More central drop and pickup points were established in the wide former forklift aisles, and no more than a day's supply of any part could be found there. With few exceptions, materials handling was done by the operators themselves because container weights were set well below OSHA and ergonomic limits.

Plant performance reflected the manufacturing concept changes. Pikeville went from being the highest-cost to lowest-cost plant in the Ferro (now JCI) group. The total number of employees decreased from 450 to 330 (attrition only), while the volume of production doubled. The manufacturing cost included less than 3 percent as a cost of labor, and lead time was measured in units-per-hour multiples. Quality soared, winning Ford's Q-1 award for the Pikeville facility.

In 1986, the Pikeville plant was selected for the JCI International Excellence Award, and one of its long-time operators went to headquarters to receive the trophy. In 1987, it appeared that a contract with Ford (Taurus/Sable) would double volume once more. So it was facilities change time again! (For more detail see Bell and Burnham's *Managing Productivity and Change* [1991, pp. 422–450].)

What does this recitation suggest? Clearly, the facilities function was being managed integratively. Product development, manufacturing engineering, plant organization and layout, and production operations all were explicitly interdependent. The site development team was the whole management organization working together.

As the manufacturing concept changed, so, too, did the way material was purchased and controlled, fabrication and assembly operations conducted, and the work force empowered to make the products. Continuous improvement was a fact of life, and total quality became a realizable goal.

Internally, the plant was modified to correspond with the new concepts. Light, floor locations, bulletin boards, meeting areas, material flow paths, ventilation, and power supplies all had to be reworked. The Engineering Department studied what equipment would be needed to assemble and test a new product, but workers had the last say on exactly how to lay out the cell. Operating maintenance was done for the most part by the operators, and quick signals were sent when anything off-spec was noted. Stringent quality checks assured that process capabilities were being maintained.

Both equipment and plant modification costs were expensed during the year they were incurred, rather than going the capital-justification route. This meant that paybacks had to be large or costs low. In fact, both occurred. Many small changes—e.g., standardized shut heights for dies, home-made reversible fixturing, the move to small containers that could be handled manually, cross-training for all personnel—were considered operating costs.

The plant's success was achieved rapidly and inexpensively, because of the teamwork developed among all of the managers, staff, supervisors, and operating personnel. The *supporting* role of the site team was pervasive, challenging, and *fun*.

REFERENCES

Bell, Robert R.; and John M. Burnham. *Managing Productivity and Change*. Cincinnati: South-Western Publishing Company, 1991.

Burnham, John M.; and Ramachandran (Nat) Natarajan. *Manufacturing Processes Student Guide*. Falls Church, Va.: APICS, 1992.

Hayes, William H.; and Steven C. Wheelwright. *Restoring Our Competitive Edge: Competing Through Manufacturing*. New York: John Wiley and Sons, 1984.

Itani, H. Kentaro. "Why Is a Systematic Approach to Integrated Facility Management Crucial to World Class Manufacturing?" Presentation. APICS National Conference, Orlando, Fla., April 1990.

James, Robert W.; and Paul A. Alcorn. *A Guide to Facilities Planning*. Englewood Cliffs, N.J.: Prentice-Hall, 1991.

Nakajima, Seiichi. *Introduction to TPM: Total Productive Maintenance*. Portland, Ore.: Productivity Press, English translation, 1988.

Tompkins, James A.; and John A. White. *Facilities Planning*. New York: John Wiley and Sons, 1984.

CHAPTER 10

INTERNAL AND EXTERNAL CUSTOMERS: SUMMARY AND FORECAST

The site development team has learned, through its ongoing involvement with the new plant, that the many customers for the facility must be fully satisfied if the project is to succeed. The challenge is for each of the elements of site operations to have these same concerns for internal customers. And the entire facility must focus on meeting the needs of the external public, whose approval is required to achieve economic viability. With the plant fully operational, many of the team members will be seeking new challenges with other site development projects.

MEETING CUSTOMERS' NEEDS

The variety of conflicts and resolution possibilities and the potential effectiveness of an ongoing site development team was detailed in Chapter 2. In Chapter 5 it was suggested that systems analysis was a useful tool for examining those other functional areas which would have a stake in the subject being considered. In this chapter, you will be given another tool: examining the situation from the standpoint of suppliers and customers. This, in effect, also summarizes the viewpoint of the text: that a facility's product must be developed integratively so that the needs of its customers and the requirements for its many suppliers can be fulfilled effectively.

Consider the typical system shown as Figure 5–7 and repeated here in Figure 10–1. Examination of the inputs suggests that manufacturing is the customer and that the input sources are the suppliers—of information,

FIGURE 10–1
Inputs for Manufacturing

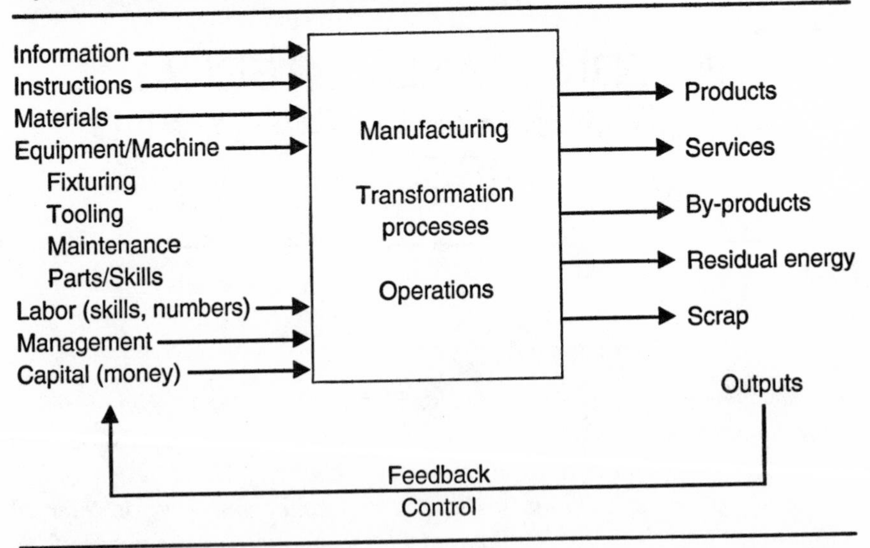

materials, and energy. Examination of the system focusing on outputs suggests that other customers exist, of course, with expectations and requirements that the manufacturing supplier must seek to fulfill.

INTERNAL AND EXTERNAL CUSTOMERS

A Supplier-Customer Relationship

The outcome of operating the system will be in doubt if needs are not clearly communicated to suppliers. Their adequacy and precision will affect supplier ability to respond. Only when both customer and supplier agree on what is needed *for the system*—and what part each must play—can progress be made.

Internal Customers

Pursuing the concept of suppliers and customers throughout the complexities of the many influences on any system will be fruitful if it leads to an understanding of the mutual benefits and a commitment to work together for continuous improvement of the system.

IFM (the site team) relates with internal elements of the firm as the processor or transformation manager. These internal elements are either customers for the IFM output or suppliers of inputs to IFM, and some of them take on different roles at different times.

These "different times" are the organization's evolving natural hierarchies, horizons spanning years strategically but only months or days when addressing specific products, processes, or facilities.

Manufacturing

As part of strategic planning, a set of key manufacturing tasks is a necessary input to the facilities strategy and to the plant charter and mission. These tasks must be expressed as explicitly as the time horizon will permit and their meaning made clear even though details are lacking. The SD team must capture and retain these manufacturing policies and seek guidance on how best to implement them at the plant.

Manufacturing operations, as part of site development, is the primary customer of IFM and will remain so for the life of the facility. It is the team process that transforms the manufacturing policy guidance into structure, machinery, layout, and materials management equipment. Further, sustainable competitive advantage requires synchronization of maintenance with production schedules and highlights the customer role for both equipment capability and availability.

Logistics

Effective logistics management and its integration into the manufacturing supply, conversion, and distribution system is a specific response to customer expectations. Capacity, locational, and facilities strategies are dependent on the ability of the logistics system to feed the supplies to the production locations and to move the products to the customer. In this context are all the aspects of outsourcing, vertical integration, and global manufacturing.

Receiving and Distribution

Ultimately, materials will come and go, whether through common physical receiving and shipping areas, or point-of-use delivery and storage, then end-of-line pickup by waiting trailers for distribution. And so these two materials specialties (receiving and shipping) will supply information in the planning mode and will be IFM customers in support of their operations.

Production Planning and Control

Equipment availability and reliability are necessary inputs to the schedule. Generally, the site team is a supplier of information to production planning and control (PPC), especially as it relates to the maintenance schedule.

Inventory must be managed in that awareness. Intertwined with receiving and distribution, IFM is a customer for information and supplies a structure and layout corresponding with PPC needs. While "operations" includes both PPC and manufacturing, manufacturing is a customer of PPC, so that the more broadly based system reflects the way the plant will operate and not only PPC's needs.

The site team is a customer for inventory services for the maintenance, repairs, and operating (MRO) stores aspects of the support responsibility, or it must provide these services for itself. Purchasing is almost certain to be involved as well.

Engineering

At a number of stages in the development of the site plan, IFM is a customer for information from engineering. Decisions on technology and process development issues, agreement on manufacturing concept, and the target quality benchmarks will greatly affect both equipment selection and maintenance schedules. IFM is also an information supplier from the same standpoints as the site plan unfolds and execution takes place, and it needs to have core engineering representation.

For the operating plant, the SD team is the focal point for both continuous improvement and grayfield modification planning. IFM is an information supplier to marketing because of possible cost savings through waste and secondary product sales, and the timing of significant modification downtimes, as these affect product output.

Equipment and Maintenance

IFM needs information from the equipment manufacturer and from other plant engineers with experience operating that model of machine. As an information customer of manufacturing, the site team seeks accurate data on equipment performance during operation. The facilities database will use this information to develop maintenance plans. The team negotiates with manufacturing for predictive and preventive maintenance plans and supplies or arranges for training for maintenance technicians as new equipment is planned or cross-training is implemented.

Work Force

All of the personnel-related facilities—bathrooms, cafeterias, meeting spaces, offices, and lockers, for example—are provided for the operations customer. Lighting, ventilation, heating and cooling, fire-fighting equipment, communications and data-gathering devices are also an IFM responsibility. How well these facilities are designed and maintained affects plant morale, safety, and productivity.

IFM is a supplier of maintenance training to equipment operators (operator-centered maintenance) through the IFM maintenance staff. The team is supplied information by equipment operators (operating maintenance carried out, equipment performance data, and improvement ideas) and thereby is a customer.

Support Staff

The mixed facility places a number of PPC, accounting, or engineering staff directly in the production areas they are to support. There may be special needs—e.g., desks, computer power, lighting, telephones, and perhaps modular workstations with sound-deadening panels—to enable these colocated staff members to support their manufacturing customer.

In the noncolocated offices, IFM has design, support, maintenance, and security responsibilities. The office "customers" are frequently in need of special services, and because of the usual personnel density, IFM will probably stay in close contact.

One of the most frequently outsourced services is office maintenance, and IFM is seen as the supplier of the service, even though an outside contractor performs it. The team is also a customer for information from both the office staff and the contractor.

Human Resources Development and Training

In common with all site personnel, IFM is a customer of Human Resources (HR) for both staff increases and transfers and for skills enhancement. On the other hand, IFM may be a supplier to HR through its maintenance technicians as trainers. In rapidly changing organizations, there may be a regular site team member from HR.

Data Processing

IFM is both a customer of and a supplier to Data Processing (DP). All maintenance productivity records will be supported by DP, and so will plantwide services such as payroll. Data on safety, equipment, hazardous materials, accidents, and regulatory compliance, as well as other history files, will be securely held in the database and kept accessible for authorized users. IFM is most often the source of such data, and so the symbiosis is clear.

Data Processing is certainly a customer for various facilities support services. Most computer rooms need filtered electric power, extra insulation, extra cooling, backup power supplies, and special lighting, in addition to their personnel needs. Because the whole site is the computer room customer and needs are dynamic, flexibility and modularity of design are highly relevant. Design requirements may include raised decking for passing cables underneath, shielded power and data cable troughs, and increasingly, the electronic equivalent of a clean room that rejects stray electromagnetic currents.

Computer rooms, local area network stations, and especially the proprietary technical and financial information often found there are primary targets of the unscrupulous. Such concentrations of equipment are at risk from natural disasters as well. Remotely located archival information files that are kept current at the site and backed up daily reflect the seriousness of the threat. Here, IFM security and disaster planning are likely to attract widespread interest.

CAD/CAM System

Regulatory compliance, physical-asset protection, and accurate information for continuous improvement and grayfield modifications necessitate having accurate, quickly retrievable IFM information. Low-cost desktop computers and user-friendly software have made electronic files more common. Network capabilities have made it feasible to hold electronic facilities files at more than one location, reducing the risk of total data loss.

Apparently, there is an emerging trend to have equipment and furniture "icons" available through the manufacturing CAD/CAM system and to have the facilities database integrated with that for products and processes. This is certainly logical, because most of the structure, layout, and material flow-pathing is of common interest to many within the plant.

External Customers

Companies able to meet—(or better, exceed!)—the expectations of their customers tend to have growing sales and profits. If the customer is a regulator, the same holds true: meet or exceed standards and the relationship will be more rewarding. IFM has many opportunities to support other company functions in this regard.

There are also areas in which *only* the site team is likely to be involved: effluent disposition, safe site access for personnel and equipment, truck routings reaching the site and on it, and a variety of subcontract professional and maintenance services.

Product Customers

Most of these external connections will be indirect, through support of manufacturing and product-process development. The site team will be guided, however by the plant charter and the inputs from marketing and manufacturing to reflect the best ways to configure the facility in response to customer needs.

Warehousing and Distribution Facilities

These specialized facilities are developed by team efforts to handle regional demand directly rather than from the plant. Just as for the production facility, these DCs need to work well with suppliers and customers and help IFM to carry out materials-handling schemes that maximize effectiveness.

Finished Goods Transporters

Common carrier rail, air, and trucking companies compete with each other and with the premium delivery services for the opportunity to handle product transportation to the marketplace. The site traffic patterns, truck dock or rail spur configuration, and the volume and frequency of shipment must lead to a close match between vehicle and facility.

Transport Companies

The transport supplier should feel a responsibility to meet customer needs, including frequency, capacity, and configuration. The responsibility of the supplier company is to arrive on time, sometimes with a specially built vehicle, to enter through the appropriate gate, and to deliver the parts and/or materials to the designated place. Configuration must suit this mutually-agreed-upon system. Site development team members must work with the PPC Department to have enough road and apron capacity to handle the resulting traffic.

By-product and Waste Processors

Many production processes dispose of heat, scrap, human and process wastewater, and a variety of other system outputs. From the IFM/SD standpoint, these outputs are as important as the primary products themselves. Essentially, IFM must design the customer interface for these in the same way that it does for the primary products or incoming materials. This customer may be a public enterprise (perhaps a water treatment plant) with regulatory and rate-setting powers.

External Suppliers

These supplier inputs to the IFM scope of responsibility are of three kinds: materials, information, and services (through people or vehicles)—in several cases, occurring in combination, which makes IFM "solutions" more complex.

Division/Corporate Facilities Guidelines

These sources supply information supplemental to the charter/mission data already available to the SD team. These suppliers should be able to respond to the Checklist questions listed at the end of Chapter 5 or obtain the information to help with planning detail.

With the outsourcing trend in facilities management (see Chapter 9), a large number of outside service suppliers may be working with IFM. The range of services makes it difficult to generalize, but access, safety, work spaces, storage, coordination and special-needs communication, contract fee arrangements, and security are all factors needing attention.

Utilities Suppliers

For water, energy, fuel, fire and police suppliers, the site team has very high coordination needs in both planning and operations phases. A typical industrial customer will have a dedicated water supply, power supply, and perhaps fuel supply (natural gas, for instance). Wastewater drain lines back to the water treatment plant may also be dedicated.

The goal is to pay one-time costs of installation and connect fees that will serve the facility for a number of years and determine how to stay within that capacity, as the suppliers may have the "whip hand." Although capacities *can* be changed, these suppliers have rate-making power or can charge for extra services, so good contingency planning can contribute to lowest life cycle cost. Further—beyond the direct fees for providing the expansion—interruption of services because of the need to enlarge pipeline capacity or change out transformers, for example, can be very costly to production.

Emergency Services

Both fire and police departments are suppliers, and the expected need for their services should be discussed well in advance of making security arrangements and disaster plans. The company or the contract security and fire protection forces should interface with their municipal opposite numbers to avoid gaps in protection. This will help obtain the best fire protection rating fee classification.

CUSTOMER EXPECTATIONS

Customer expectations have an indirect and apparently infrequently considered influence on manufacturing. Marketing and sales professionals and customer service people are quite aware of what the customer is saying, but filtering sometimes makes it difficult to hear. However, some specific plant level activities have the goal of learning about customer expectations and developing effective responses. It may not be possible to directly match every customer's need, and some trade-offs are inevitable. But competitive benchmarking can serve for both communications and improvement targeting, with a variety of responses in purchasing, design, and production. Figure 10–2 summarizes these many interactions and customer relationships.

Examples

At Xerox in 1979, the decline in market share was painfully evident. Taking advantage of global affiliations and manufacturing in Europe and the Far East, Xerox began to take the measure of competitive details. The best competitor's supplier quality was 10 times better than that of suppliers serving Xerox. Delivered-product defects were an order of magnitude less among top global competitors. Xerox lead times were long, inventory was high, and product costs were far out of line, according to the benchmarks. The denial syndrome was the immediate reaction at Xerox. But what followed was a commitment to becoming the benchmark that competitors would use to describe quality. Targets were set to record and track the gap and to close it.

The questions asked by those in the various aspects of manufacturing were "How will this gap affect what I must do to help the company?" and "How does the likelihood that will change due to competitive pressure and affect what we do at this site?" These questions led to more-detailed discussions about expansion, equipment upgrades, staffing additions, and altering support systems and procedures. They can provide sharp focus on improvement areas in light of the competitive move, or shift in power, or change in consumer expectation. And Xerox heeded the message.

Time-Based Competition

Such time-related aspects as new-product development time (first-to-market goal), manufacturing lead time, flexibility, and order-to-ship time become as important to success as product quality, inventory turns, and manufactured cost. Gains by competitors or the entry of a new player into the manufacturing game should occasion a number of what-if analyses. A kind of benchmark-gap analysis becomes a motivator for change.

Design lead times for automobiles in Japan are barely half those in North America, and they are achieved by the thoroughly integrated efforts of multifunctional teams and excellent manufacturing execution. The result is a first-to-market competitive advantage.

Logistics systems can quickly bring products to market and support them as needed. Their availability and responsiveness enabled Caterpillar to deliver replacement parts worldwide within 48 hours. By using a wide range of delivery patterns, Hewlett-Packard maintains a response time of no longer than four days for its peripheral products.

The intelligent exchange of information can gain precious time in responding to customer needs. The Westinghouse plant at Asheville, North Carolina, noted that orders taken by a salesperson could be *on the factory floor* eight minutes later with a confirmed shipping date.

Internal to manufacturing, modular design allows "plain-vanilla" subassemblies that can be combined to meet unique final configurations. Hewlett-Packard has used this logic to handle the many millions of possible end item configurations for its 9000 Series computers at Fort Collins, Colorado. Dana does this in its European OEM service regions, which fabricate automobile parts and subassemblies. JIT/TQC develops synchronous flow that minimizes inventory and lead times, while providing prompt feedback throughout the system.

Benneton, the well-known fashion merchandiser, used information technology to predict fashion trends and at the same time cut down on

FIGURE 10–2
External and Internal Relationships

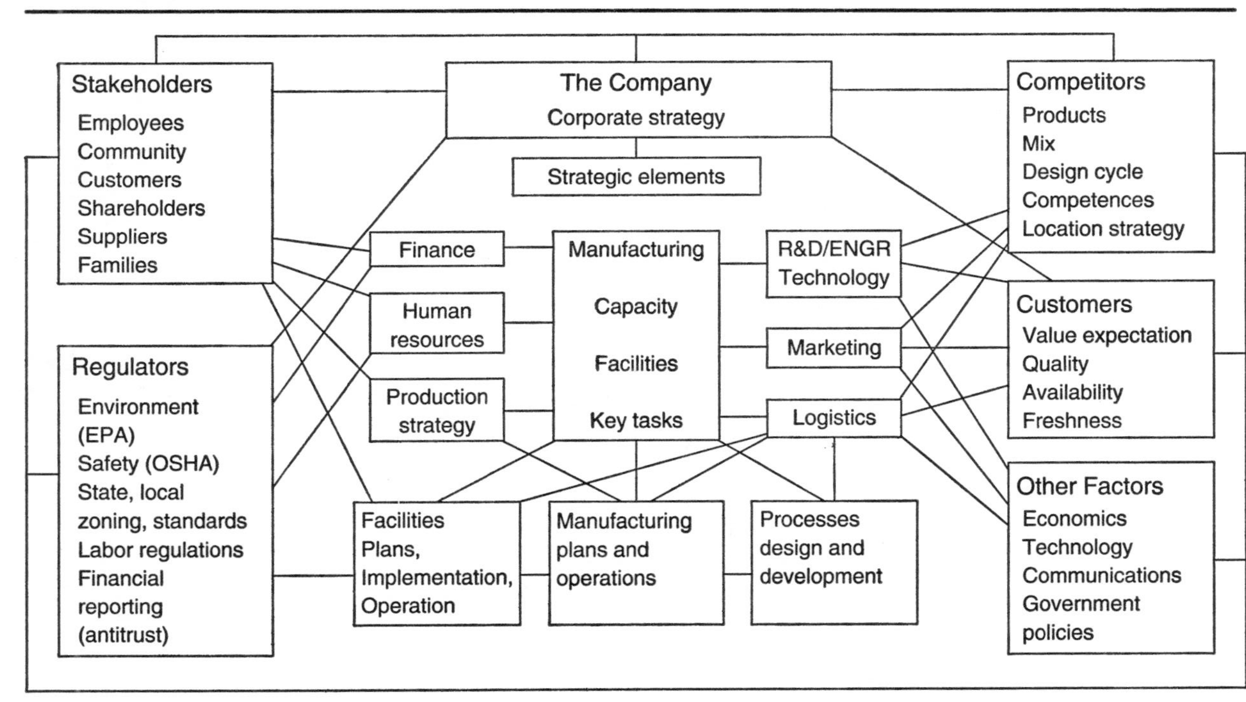

Source: John M. Burnham and Ramachandran (Nat) Natarajan, *Manufacturing Processes,* Student Guide (Falls Church, Va.: APICS, 1992), Fig. 9–2, p. 9–4. Reprinted with the permission of the publisher.

inventory. By tracking sales of styles and sizes, precise replacement merchandise was quickly assembled by a network of Italian dressmakers. Dyeing took place *after* assembly so that color obsolescence was avoided. This enabled Benneton to stay ahead of Pacific Rim competitors that have long lead times and compete on price alone.

Product Attributes

Customer expectations must be integrated with the product attributes developed during design. Style, appearance, features, quality, and durability are all established fairly early. Eighty-five percent of manufacturing cost is determined in the initial phases of design. The SE product/process development team must know the targets and address the whole range of issues concurrently.

Product Mix, Variety, and Life Cycle

At least in the short run, shifts in mix will affect capacity, timeliness, and cost. Variety that is planned for and matched with an appropriate manufacturing concept can enhance competitive advantage. Capacity and ease-of-make are dependent on the product design and the mix assigned to the manufacturing plant. External forces—sales results, for example—can have significant short-term impact on all aspects of manufacturing. Short product life cycles are the norm in many businesses and have considerable influence on the choices of equipment, fixturing, tooling, and layout.

Avoid Closed Systems

Note how the other areas of the manufacturing processes CIRM module and those of products and services and logistics bear directly on the capacity, location, technology, structure, layout, materials handling, scheduling practices, and staffing at the site.

Global competition sets benchmarks for quality. Quality function deployment (QFD) outlines the customer's design mandates. Products, parts, processes, and controls need to match customers' expectations. Internally, the manufacturing system capability is defined, and operations and process choices follow. Designed experiments are used to optimize product and process designs. Trade-offs throughout the system help match with minimum loss the target costs and quality.

The decision to add or defer capacity increments is strategic, driven by known industry production capability, known or suspected competitor capacity additions, and projections about the economy and product demand. This applies equally internal suppliers—(make)—and to the supplier network—(buy, or outsource). Thus, it is industry capacity and a company's desired positioning which direct internal and external adjustments.

Flexibility and Responsiveness

Depending on the competitive strategy, plants may be quite rigid or quite responsive and flexible. Capacity, the manufacturing concept, and equipment selection can enable manufacturing to respond easily. Cross-training, efforts toward quick changeover, and excess capacity can all help with flexibility.

If diversity is limited, repetitive manufacturing becomes feasible. Management can focus on a set of coherent, well-defined manufacturing tasks and achieve excellent performance. The example of Nucor shows the benefits of this kind of focus. However, within such plants, there is little flexibility. Lines are completely dedicated to their defined product family.

Continuous Improvement

All competitors have their own experience curves, and each is seeking advantage. Gains are much more than simply direct labor hour reduction. The product/process development activities can be greatly improved over time as teams learn focused integrative thinking. Commitment to continuous improvement and a clear understanding of the key tasks for manufacturing provide for the most effective improvement activities.

Communications

The need to perfect communications both technically and in terms of their content has never been greater. Within the company and its organizational elements, and across the boundaries to reach supplier partners and customers, electronic data interchange has provided tracking ability, and bar coding has practically eliminated error in description. But asking the *right* questions and understanding the responses is still a major challenge. Customer communications are critical because they form the fundamental means for responding to expectations.

Recently, a small but specialized supplier of industrial duct heaters, concerned about the manufacturing error rate for its custom product line, sought advice about applying statistical process control (SPC) techniques to the engineering design activities generating the manufacturing order for the shop. Investigation revealed that the more likely area for study was the source of the order: a distributor network and manufacturers' representatives who served many different suppliers and customers. The terms in which the orders were expressed were not sufficiently clear to make the design process error free.

Internal Effects

Detailed analysis of the internal effects of shifts in the markets (and the characteristics of the customers), the competitors, and the suppliers goes far beyond the scope of this book. A checklist included as an appendix to this chapter summarizes the concerns that IFM and the site development team need to have.

LESSONS FOR THE INTEGRATED RESOURCES MANAGER

This final chapter has moved across the range of both integrative facilities management and customer-supplier relationships inherent in the entire CIRM series. The goal has been to invite each reader-manager to carry out a similar (and sufficiently exhaustive) study when faced with a situation requiring analysis or decision. This chapter offers a point of view, rather than the many why-to, what-to, and how-to details of the previous chapters.

The lesson is a simple one: The supplier-customer interdependency is profound, and there are many more legitimate suppliers and customers than traditional wisdom has acknowledged. Twenty or more years is a very long time to have to live with mistakes caused by failure to develop and use the valuable information sources and professional and personal relationships that are available to help with the integrative facilities management process.

APPENDIX

TABLE 10A–1
Internal Change Sources

Financial (capital, cost, projected ROE, ROA)
R&D/engineering, technology
Accelerating or delaying capacity/facility actions
Product, process, or production decisions
Extending, modifying, terminating, increasing, reducing volume
Changing product or facility to meet benchmark mandates
Logistics system development—suppliers, distribution

Procedural (software) or personnel changes
  Customer- or competition-driven
  Other stakeholder influences

Major conservation-driven efforts/recycling

TABLE 10A–2
External Change Sources

Voice of the customer
Time-based competition
Quality function deployment
"Green" engineering/environmental pressures (aerosol, recycling) as customer-driven
Logistics, supplier-to-ultimate-customer system (time, value)

Effects
  Required system modifications to retain customers, competitive positioning.

Responses
  Ongoing study of customer and expectations
  Contingency planning
  Tradeoffs
  R&D, engineering, technology efforts
  "Systems" work (logistics)
  Product or facility changes to meet new requirements
  Human resources development, training, involvement
  "Green" engineering
  Greenfield or grayfield facility work

Source: (Tables 10A–1, 10A–2) John M. Burnham and Ramachandran (Nat) Natarajan, *Manufacturing Processes,* Student Guide (Falls Church, Va.: APICS, 1992), Figs. 9–5, 9–9. Used with the permission of the publisher.

INDEX

D

E

F

G

H

I

P

About APICS

APICS, the educational society for resource management, offers the resources professionals need to succeed in the manufacturing community. With more than 35 years of experience, 70,000 members, and 260 local chapters, APICS is recognized worldwide for setting the standards for professional education. The society offers a full range of courses, conferences, educational programs, certification processes, and materials developed under the direction of industry experts.

APICS offers everything members need to enhance their careers and increase their professional value. Benefits include:

- Two internationally recognized educational certification processes—Certified in Production and Inventory Management (CPIM) and Certified in Integrated Resource Management (CIRM), which provide immediate recognition in the field and enhance members' work-related knowledge and skills. The CPIM process focuses on depth of knowledge in the core areas of production and inventory management, while the CIRM process supplies a breadth of knowledge in 13 functional areas of the business enterprise.
- The APICS Educational Materials Catalog—a handy collection of courses, proceedings, reprints, training materials, videos, software, and books written by industry experts . . . many of which are available to members at substantial discounts.
- *APICS The Performance Advantage*—a monthly magazine that focuses on improving competitiveness, quality, and productivity.
- Specific industry groups (SIGs)—suborganizations that develop educational programs, offer accompanying materials, and provide valuable networking opportunities.
- A multitude of educational workshops, employment referral, insurance, a retirement plan, and more.

To join APICS, or for complete information on the many benefits and services of APICS membership, **call 1-800-444-2742 or 703-237-8344.** Use extension 297.